Illuminate Publishing

T0173278

# Eduqas
# **Physics**
## A Level Year 1 & AS

## *Study and Revision Guide*

Gareth Kelly
Nigel Wood
Iestyn Morris

ip

Published in 2016 by Illuminate Publishing Ltd, P.O Box 1160,
Cheltenham, Gloucestershire GL50 9RW

Orders: Please visit www.illuminatepublishing.com
or email sales@illuminatepublishing.com

British Library Cataloguing in Publication Data

A catalogue record for this book is available from the British Library
ISBN 978-1-908682-72-7

Printed by Severn, Gloucester

11.20

The publisher's policy is to use papers that are natural, renewable and recyclable
products made from wood grown in sustainable forests. The logging and manufacturing
processes are expected to conform to the environmental regulations of the country of
origin.

Every effort has been made to contact copyright holders of material reproduced in this
book. If notified, the publishers will be pleased to rectify any errors or omissions at the
earliest opportunity.

This material has been endorsed by Eduqas and offers high quality support for the
delivery of Eduqas qualifications. While this material has been through a Eduqas quality
assurance process, all responsibility for the content remains with the publisher.

WJEC examination questions are reproduced by permission from WJEC

Editor: Geoff Tuttle
Cover and text design: Nigel Harriss
Text and layout: Neil Sutton, Cambridge Design Consultants

## Acknowledgements

We are very grateful to the team at Illuminate Publishing for their professionalism,
support and guidance throughout this project. It has been a pleasure to work so closely with
them.

# Contents

# How to use this book

As examiners and former examiners we have written this new study guide to help you be aware of what is required for – and structured the content to guide you towards – success in the Eduqas AS examination, or your year 12 studies towards an A level in Physics.

If you are a student following the full A Level course using this book, you will find that the AS specification Sections match the A Level specification Sections as follows:

AS      C1.1 C1.2 C1.3 C1.4 C1.5 C1.6 C1.7 C2.1 C2.2 C2.3 C2.4 C2.5 C2.6 C2.7 C2.8

A Level  C1.1 C1.2 C1.3 C1.4 C2.5 C2.7 C3.7 C2.1 C2.2 C2.3 C3.1 C3.2 C3.3 C3.4 C3.5

There are two main sections to the book:

## Knowledge and Understanding

This first section covers the key knowledge required for the examination.

You'll find notes on the content of both examination components:

- Component 1 Motion, Energy and Matter
- Component 2 Electricity and Light

including the practical and data-handling skills which you will need to develop.

In addition there are a number of features throughout this section that will give you additional help and advice as you develop your work:

Component introduction

The key sub-sections are listed with their page references and their corresponding exam questions. Each then has a short summary giving you an essential overview of the area of study, plus a revision checklist as you work through your revision process.

- **Key terms**: many of the terms in the Eduqas specification can be used as the basis of a question, so we have highlighted those terms and offered definitions.
- **Quickfire questions**: are designed to test your knowledge and understanding of the material.
- **Pointer and Grade Boost**: offer extra examination advice based on experience of what candidates need to do to attain the highest grades.
- **Extra questions**: feature at the end of each area of study providing you with further practice at answering questions with a range of difficulty.

# Exam Practice and Technique

The second section of the book covers the key skills for examination success and offers you examples based on suggested model answers to possible examination questions. First, you will be guided into an understanding of how the examination system works, an explanation of Assessment Objectives and how to interpret the wording of examination questions and what they mean in terms of exam answers.

A variety of structured and QER practice questions is provided taken from across the AS specification and model answers given. This is followed by a selection of examination and specimen questions with actual student responses. These offer a guide as to the standard that is required, and the commentary will explain why the responses gained the marks that they did.

Most important of all, we advise you to take responsibility for your own learning and not rely on your teachers to give you notes or tell you how to gain the grades that you require. You should look for extra reading and additional notes to support your study in physics. It's a good idea to check the awarding body website – www.eduqas.co.uk – where you can find the full subject specification, specimen examination papers, mark schemes and in due course examiner reports on past years' exams.

Good luck with your revision!

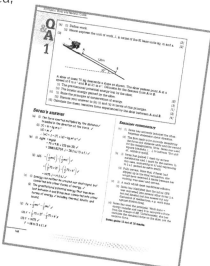

# Component 1

# Knowledge and Understanding

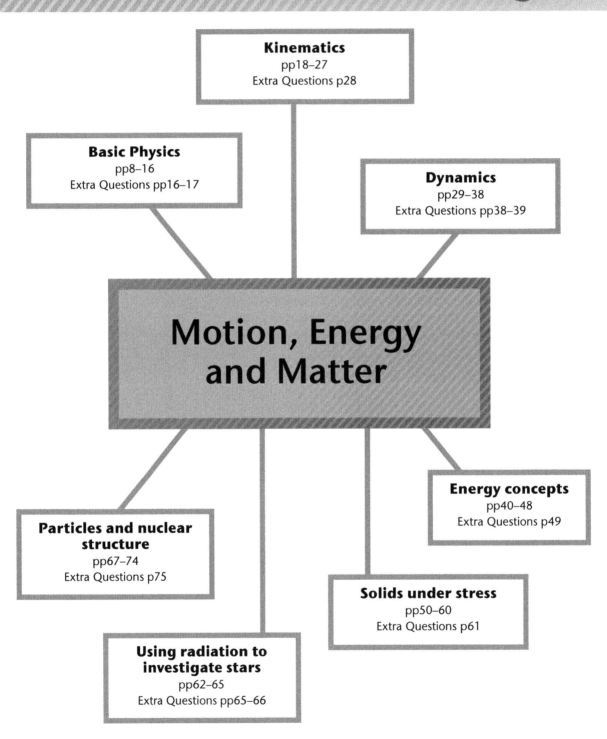

**Kinematics**
pp18–27
Extra Questions p28

**Basic Physics**
pp8–16
Extra Questions pp16–17

**Dynamics**
pp29–38
Extra Questions pp38–39

## Motion, Energy and Matter

**Energy concepts**
pp40–48
Extra Questions p49

**Particles and nuclear structure**
pp67–74
Extra Questions p75

**Solids under stress**
pp50–60
Extra Questions p61

**Using radiation to investigate stars**
pp62–65
Extra Questions pp65–66

## Basic physics

Units, dimensions, basic ideas on scalar and vector quantities and the differences between them; the ideas and skills you need to progress to further study of Newtonian mechanics, kinetic theory and thermal physics.

p8–16

## Kinematics

Rectilinear and projectile motion; the study of accelerated motion in a straight line; the motion of bodies falling in a gravitational field; the independence of vertical and horizontal motion of a body moving freely under gravity.

p18–27

## Dynamics

The concept of force and free body diagrams; Newton's laws of motion and the concept of linear momentum; the principle of conservation of momentum is used to solve problems involving both elastic and inelastic collisions.

p29–38

## Energy concepts

The relationship between work, energy and power; the principle of conservation of energy, and the link between work and energy via the work–energy relationship.

p40–48

## Solids under stress

The behaviour of different kinds of solids under stress; the concepts of stress, strain and the Young modulus; the work done deforming a solid is related to the strain energy stored; comparing the behaviour of metals, brittle materials and rubber under stress.

p50–60

## Using radiation to investigate stars

The continuous emission and line absorption spectra of the Sun; the use of the Wien displacement law, Stefan's law, and the inverse square law to investigate properties of stars, including luminosity, size, temperature and distance.

p62–65

## Particles and nuclear structure

Matter is composed of quarks and leptons; the quark composition of the neutron and the proton and the idea that quarks and antiquarks are never observed in isolation; the four interactions experienced by particles; the application of the conservation of charge, lepton number and quark number to particle interactions.

p67–74

Basic notes   Good grasp   Fully revised

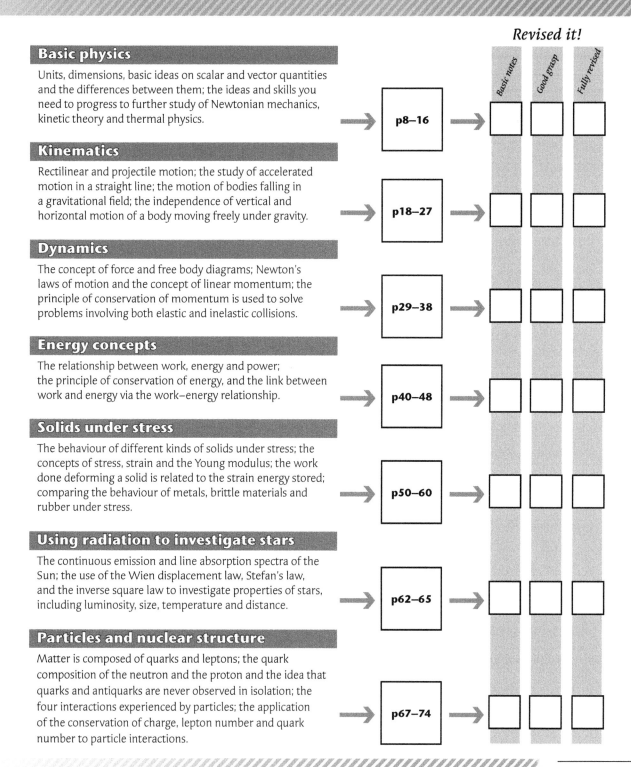

# 1.1 Basic physics

Despite its somewhat uninspiring title, this section of the book is vital for successful study of physics. Many of the concepts are familiar from GCSE so you should treat this as a revision section. We start off with looking at units. You cannot communicate anything in physics without understanding units.

## Rules for handling units

1. Two quantities can only be equal if they can be expressed in the same units, e.g. 1 s can never be equal to 1 cm.
2. You can only add or subtract two quantities if you can express them in the same units, e.g. 1 s + 1 cm is nonsense, but:

   1.00 m + 1 cm = 1.01 m (or 101 cm)
3. If you combine two quantities by multiplying (or dividing), the units are multiplied (or divided), e.g.

   $$\text{unit of } v = \frac{\text{unit of } x}{\text{unit of } t}$$

   $$= \frac{m}{s} = m\ s^{-1}$$

## » Pointer

Use square brackets [..] as a shorthand to indicate 'the unit of'.
So [F] = the unit of F (force) = N and

$$[v] = \frac{[x]}{[t]} = \frac{m}{s} = m\ s^{-1}$$

## » Pointer

In the equation $v^2 = u^2 + 2ax$, the 2 has no units because it is just a number. So the unit of $2ax$ is the same as the unit of $ax$.

# 1.1.1 Units and dimensions

## (a) Base SI units

There are six base units that you'll need to know for this course and these relate to the most fundamental and basic properties of physics:

| Base quantity | Unit | Symbol |
|---|---|---|
| Mass | kilogram | kg |
| Length | metre | m |
| Time | second | s |
| Electric current | ampère | A |
| Temperature | kelvin | K |
| Amount of substance | mole | mol |

At this stage, you only need to know the first four units and quantities which need no introduction as you've been using them for years. You'll encounter the kelvin in the AS course and the mole in the full A level.

Look at the rules for handling units. We'll do a couple of examples to see how to apply them. The most common question asked relating to base units is checking that both sides of an equation have the same units – sometimes referred to as checking equations for homogeneity.

**Example**

Check that the equation $v^2 = u^2 + 2ax$ is homogeneous

**Answer**

Right-hand side (RHS):  $[u^2] = (m\ s^{-1})^2 = m^2\ s^{-2}$ ;

$$[2ax] = [2][a][x] = (m\ s^{-2})\ m = m^2\ s^{-2}$$

unit of acceleration

So the two terms on the RHS are equal, so

- They can be added together, and
- The unit of their sum is also m² s⁻²

Left-hand side (LHS):  $[v^2] = (m\ s^{-1})^2 = m^2\ s^{-2}$

So the units of the LHS and RHS are the same, i.e. the equation is homogeneous.

**Hint:** Set your answer out in this way: do one side (start with the complicated side), then the other side, then make a comment. QED

Another type of question is expressing SI units in terms of the base units (kg, m, s....).

**Example**

Express the units of force, the newton (N), in terms of base SI units.

**Answer**

Start with the equation:  $F = ma, \therefore [F] = [m] \times [a]$

$\therefore$ N = kg $\times$ m s$^{-2}$ = kg m s$^{-2}$.

We can take the result for the newton and go on to use the equations

$W = Fx$ and $P = \dfrac{W}{t}$ to express the joule (J) and watt (W) in terms of the base SI units.

## (b) Writing SI units – negative indices and SI multipliers

Note the following:

- At AS level and above we move on from the GCSE style of writing units: we write m s$^{-1}$ instead of m/s. Note the gap between the m and the s: ms$^{-1}$ means *per millisecond*.

- In exams, you will have a Data Booklet which contains a list of SI multipliers from a (atto $-10^{-18}$) to Z (zetta $- 10^{21}$). See Table 1.1.1. Exam questions can be set using them, e.g. a force of 36 MN. It is standard practice to keep the number in front of the multiplier to between 1 and 1000.

- Standard form is also used, e.g. $c = 3.00 \times 10^8$ m s$^{-1}$.

| Symbol | Multiple | Symbol | Multiple |
|--------|----------|--------|----------|
| a | $10^{-18}$ | k | $10^3$ |
| f | $10^{-15}$ | M | $10^6$ |
| p | $10^{-12}$ | G | $10^9$ |
| n | $10^{-9}$ | T | $10^{12}$ |
| μ | $10^{-6}$ | P | $10^{15}$ |
| m | $10^{-3}$ | E | $10^{18}$ |
| c | $10^{-2}$ | Z | $10^{21}$ |

*Table 1.1.1 SI multipliers*

**Example**

A force of 160 kN is applied over an area of 2 cm × 2 cm. Calculate the pressure and express your answer using (a) standard form and (b) an SI multiplier.

**Answer**

The area = $(2 \times 10^{-2}$ m$)^2 = 4 \times 10^{-4}$ m$^2$

(a) Pressure = $\dfrac{\text{Force}}{\text{area}} = \dfrac{160 \times 10^3 \text{ N}}{4 \times 10^{-4} \text{ m}^2} = 4 \times 10^8$ Pa (see Grade boost)

(b) $4 \times 10^8$ Pa = $400 \times 10^6$ Pa = 400 MPa.

**quickfire**

① Show that the equation $x = ut + \dfrac{1}{2} at^2$ is homogeneous.

**quickfire**

② Express the unit of pressure, the pascal (Pa) in terms of the base SI units.

**quickfire**

③ (a) Show that $J = $ kg m$^2$ s$^{-2}$.

   (b) Express the watt (W) in terms of the base SI units.

**Grade boost**

Learning the SI base-unit expressions for the newton, joule and watt can save time in an exam.

**quickfire**

④ The air resistance force, $F$, on a car is given by the equation $F = kv^2$. Find a unit for $k$.

**Grade boost**

When entering 160 kN into your calculator, put 160 × 10$^3$ rather than converting to 1.6 × 10$^5$ in your head. Let the calculator take the strain!

## Key Terms

A **scalar** is a quantity which has magnitude only.

A **vector** is a quantity which has magnitude and direction.

### quickfire

⑤ Express $5.6 \times 10^{-5}$ m using an SI multiplier.

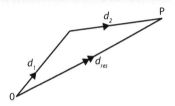

Fig.1.1.1 Resultant vector

### quickfire

⑥ Draw the resultant force, $F_{res}$.

### » Pointer

Many people find it helpful to draw a parallelogram (in this case a rectangle) of the vectors with the resultant as the diagonal. The answer is the same!

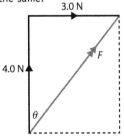

Fig. 1.1.4 Adding vectors by a rectangle.

# 1.1.2 Scalars and vectors

All things that you measure in physics are classified as either **scalar** or **vector**. The essential difference between a scalar and a vector is that a **vector has direction**. This means that, whereas adding two scalars is easy, you must use geometry to add two vectors.

Mass is a scalar, so $\qquad$ 2 kg + 5 kg = 7 kg

but force is a vector, hence, $\quad$ 2N + 7N = ... depends on their directions.

You need to be able to add and subtract vectors. Fig. 1.1.1 shows how to add two displacements $d_1$ and $d_2$. Suppose a car starts from O and has a displacement $d_1$ followed by a displacement $d_2$. The car ends up at P with a displacement of $d_{res}$. What you've done is to add the two vectors $d_1$ and $d_2$, giving a **resultant** displacement of $d_{res}$. All vectors add in this way – you do the same for forces, velocities, accelerations etc. (Note that it's good practice to give the resultant a double arrow.)

Here is a list of most of the scalars and vectors you'll meet at AS level:

Scalars – density, mass, volume, distance, length, work, energy (all forms), power, charge, time, resistance, temperature, pd, activity, pressure.

Vectors – displacement, velocity, acceleration, force, momentum.

## (a) Adding vectors

You can use scale diagrams to find the resultant vector ($v_1 + v_2$). You can also be asked to calculate the resultant of two vectors at right angles using simple trigonometry (trig), and then use the answer to do a further calculation.

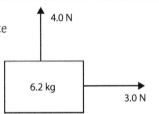

Fig. 1.1.2 Resultant force

**Example**

Find the resultant force on the box in Fig. 1.1.2 and use it to find the acceleration.

**Answer**

We'll use Pythagoras' theorem to find the resultant force: $F^2 = 3.0^2 + 4.0^2 = 25$

So $F = \sqrt{25} = 5$ N

Is this the final answer?

No! A force is a vector, so you need to say what its direction is. That's what that sneaky $\theta$ is for in Fig. 1.1.3.

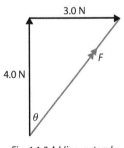

Fig. 1.1.3 Adding vectors by Pythagoras' theorem

$\tan \theta = \dfrac{3.0}{4.0} = 0.75 \quad$ So $\theta = \tan^{-1} 0.75 = 36.9°$.

∴ The resultant force is 5.0 N at 36.9° to the 4.0 N force.

∴ Acceleration $= \dfrac{F_{res}}{m} = \dfrac{5.0 \text{ N}}{6.2 \text{ kg}} = 0.81$ m s$^{-1}$ at 36.9° to the 4.0 N force.

## (b) Subtracting vectors

We subtract vectors in the same way as adding them if we remember that $a - b$ is the same as $a + (-b)$. Fig. 1.1.5 shows what we mean by $-v$, where $v$ is a vector.....

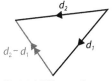

.....and Fig. 1.1.6 shows how we use this idea to subtract the vector $d_2$ from $d_1$. Subtracting vectors is important in, for example, calculating acceleration:

$$a = \frac{v - u}{t}.$$

Fig.1.1.6 Difference between two vectors

## (c) Resolving vectors

Unfortunately, vectors don't usually have the direction that you want them to have. Take the example in Fig. 1.1.7:

As the block slides down the slope, the force of gravity (weight) acts downwards but the block must move along the slope, which is in a different direction. You need to calculate the part of the gravitational force (weight) that acts in the direction of the slope. This is very similar to calculating the resultant of two vectors but in reverse. Fig. 1.1.8 gives the idea:

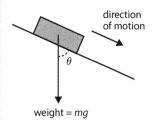

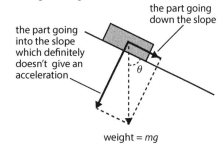

Fig. 1.1.7 Object on a slope

Fig. 1.1.8 Components of the weight

The two bold vectors have been drawn to give a resultant vector which is the weight. The clever bit about this process is that we can use the bit going down the slope to calculate an acceleration (the other bit just pushes the block into the slope).

The correct terminology for this process is resolving the vector (weight) into two components (down the slope and into the slope).

Using the trig that we know:

$$F_{\text{down the slope}} = mg \cos \theta$$

So, in the absence of friction, the acceleration of the block is $g \cos \theta$.

### Example

Calculate the horizontal and vertical components of the 500 N force on the sledge.

### ≫ Pointer

The change in a quantity, $q$, from $q_1$ to $q_2$ is $q_2 - q_1$, whether $q$ is a vector or a scalar. As a shorthand we write this as $\Delta q$ (and pronounce it 'delta $q$').

Fig 1.1.5 Vectors v and –v

### quickᴨιre

⑦ For the forces in QF6, draw the triangle to represent $F_1 - F_2$.

### quickᴨιre

⑧ A light aircraft flying at 40 m s⁻¹ due North, alters course and speed to 30 m s⁻¹ due East. Find the change in velocity.

### ≫ Grade boost

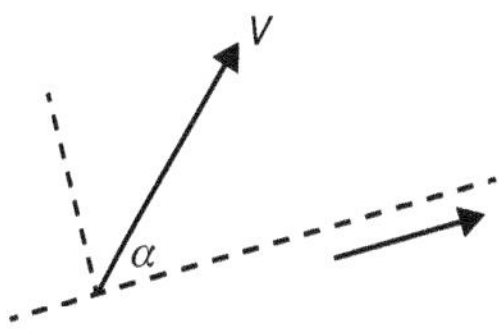

The component of $V$ in the direction of the arrow is $V \cos \alpha$. The component at right angles is $V \sin \alpha$.

⑨ The vertical component of the helicopter's thrust is equal to the weight. Calculate the thrust.

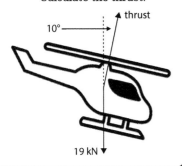

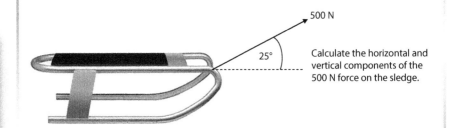

Calculate the horizontal and vertical components of the 500 N force on the sledge.

**Answer**

Horizontal component = 500 cos 25° = 453 N

Vertical component = 500 sin 25° = 211 N

**Note:** In the same way, a missile travelling at 500 m s$^{-1}$ at 25° to the horizontal has a horizontal velocity of 453 m s$^{-1}$ and a vertical velocity of 211 m s$^{-1}$.

⑩ Calculate the helicopter's horizontal acceleration. [Hint: you'll need to find its mass first.]

**Key Term**

The **density**, $\rho$, of a material is defined by

$$\rho = \frac{m}{V}$$

where $m$ is its mass and $V$ is its volume.

Unit: kg m$^{-3}$.

 **Grade boost**

Some unit conversions to learn

| | |
|---|---|
| 1 cm$^2$ | = (10$^{-2}$ m)$^2$ = 10$^{-4}$ m$^2$ |
| 1 cm$^3$ | = (10$^{-2}$ m)$^3$ = 10$^{-6}$ m$^3$ |
| 1 mm$^2$ | = (10$^{-3}$ m)$^2$ = 10$^{-6}$ m$^2$ |
| 1 mm$^3$ | = (10$^{-3}$ m)$^3$ = 10$^{-9}$ m$^3$ |
| 1 g cm$^{-3}$ | = 1000 kg m$^{-3}$ |

⑪ An empty shipping container has lengths 6.1 m × 2.44 m × 2.59 m. Calculate the mass of air it contains.

# 1.1.3 Density

This is the property of a material that gives us the concentration of its mass. The **density** of a material is a constant and does not depend on the size and shape of the sample of material. A few examples are given in Table 1.1.2

The SI unit of density is kg m$^{-3}$, so its value expresses the mass of a cubic metre of material. Thus 1 m$^3$ of gold has a mass of 19 300 kg or 19.3 tonnes (1 tonne = 1000 kg) and 1 m$^3$ of water weighs in at 1 tonne.

| Material | density / kg m$^{-3}$ |
|---|---|
| Air | 1.23 |
| Water | 1000 |
| Iron | 7 700 |
| Gold | 19 300 |
| Platinum | 21 450 |

Table 1.1.2 Densities

Another common unit for density is g cm$^{-3}$ and examiners really like getting you to convert between one unit and the other. The easiest way is as follows, taking gold as the example:

Density of gold = 19 300 kg m$^{-3}$

= 19 300 (1000 g) × (100 cm)$^{-3}$

= $\dfrac{19\ 300 \times 1000\ \text{g}}{100\ \text{cm} \times 100\ \text{cm} \times 100\ \text{cm}}$

= 19.3 g cm$^{-3}$

Because we often measure masses in g and physical dimensions in cm and mm, lots of examination questions about density (and resistivity and tensile stress) involve the conversion between these units and the basic SI units. See the Grade boost for some unit conversions.

**Example**

A student places an empty measuring cylinder on an electronic balance and presses the tare (zero) button. The reading with 80.0 cm$^3$ of carbon tetrachloride (CCl$_4$) in the measuring cylinder is 126.53 g. Calculate the density of CCl$_4$, expressing your answer in kg m$^{-3}$ to an appropriate number of significant figures.

**Answer**

Either: $\rho = \dfrac{126.53\ \text{g}}{80.0\ \text{cm}^3} = 1.58\ \text{g cm}^{-3} = 1580\ \text{kg m}^{-3}$ (3 s.f.)

or: $\rho = \dfrac{0.12653\ \text{kg}}{80.0 \times 10^{-6}\ \text{m}^3} = 1580\ \text{kg m}^{-3}$ (3 s.f.)

**Example**

Calculate the mass of a platinum wire of length 6.5 m and diameter 0.254 mm.

**Answer**

We'll need to convert some of these units. Putting everything into cm:

$\rho = \dfrac{m}{V} \therefore m = \rho V = 21.45\ \text{g cm}^{-3} \times \pi \left(\dfrac{0.0254\ \text{cm}}{2}\right)^2 \times 650\ \text{cm} = 7.1\ \text{g}$

**Pointer**

In the example, the volume of the liquid is given to 3 s.f. so the density should not be stated to more than this.

**Grade boost**

For flexibility, try the platinum wire example using kg and m as the units.

# 1.1.4 Moment – the turning effect of a force

## (a) Definition of moment

Forces can make objects rotate as well as accelerate. The turning effect of a force depends not only on its magnitude but also its distance from the point of rotation – using a long spanner you can undo a nut which is impossible to shift using a short one.

The definition of the moment of a force about a point is a bit of a mouthful (see Key terms). It boils down to this:

Moment of $F$ about **P** = $Fd$

For example, with $F$ = 35 N and $d$ = 15 cm,

Moment = 35 N × 0.15 m (note the change of unit) = 5.25 N m

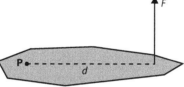

Fig. 1.1.9 Moment of a force

If the force is at an angle as in Fig. 1.1.10, you need to apply a bit of trig:

This time we need to work out the perpendicular distance.

perp dist = 0.82 sin 60° m

$\therefore$ moment = 1.5 kN × 0.82 sin 60° m

= 1.1 kN m (1100 N m)

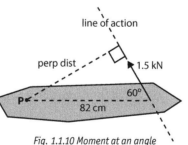

Fig. 1.1.10 Moment at an angle

**Key Terms**

The **moment** of a force about a point is the product of the force and the perpendicular distance from the point to the line of action of the force.

Unit: N m (or N cm, kN m ....)

The **principle of moments** states that for a body to be in equilibrium under the action of a number of forces, the resultant moment about any point is zero.

## (b) Principle of moments (PoM)

For an object to be in equilibrium (i.e. not changing its motion in any way) we know that:

- The resultant force on it must be zero.

There is a second useful condition concerning rotations:

- The resultant moment about any point must be zero (the **principle of moments**).

**Pointer**

Another way of stating the principle of moments is to say that the sum of the anticlockwise moments (about any point) is equal to the sum of the clockwise moments (about the same point).

## Grade boost

Conditions for equilibrium:
1. Resultant force = 0
2. Resultant moment (about any point) = 0

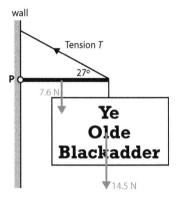

Fig. 1.1.11 Baldric's problem

### Symbols

The symbol Σ is used to stand for 'the sum of'. So ΣF means the sum of the forces (i.e. the resultant force) and ΣCM means the sum of the clockwise moments.

## ≫ Pointer

Notice that answers to the example make the resultant force on the springboard zero. This must be the case if the springboard is to be in equilibrium. We could have used this fact to calculate $F_1$ once we had found $F_2$.

## quickfire

⑫ Solve the springboard example by:
1. Taking moments about the centre.
2. Taking moments about the right-hand end.
3. Solving the resulting simultaneous equations.

---

Taken together these two conditions will give you enough information to solve quite tricky questions about the forces on an object. We'll have a look at a couple of examples, one of which has no pivot.

## Example 1

The inn sign In Fig. 1.1.11 is suspended from a uniform metal bar which is smoothly pivoted at **P**. Calculate the tension, $T$, in the light supporting wire.

### Answer

Let the length of the metal bar be $2x$. Anticipating the work on centres of gravity in the next section, we can consider the weight of the bar to act at its mid-point, i.e. a distance $x$ from **P**. Both $2x$ and $x$ are perpendicular to the lines of action of the forces and these forces have clockwise moments about **P**.

The perpendicular distance of $T$ from **P** = $2x \sin 27°$

Then, applying the principle of moments about **P**,

sum of anticlockwise moments = sum of clockwise moments

$$\therefore\ T \times 2x \sin 27° = 7.6x + 14.5 \times 2x$$

Cancelling by $x$ and rearranging gives: $T = \dfrac{7.6 + 14.5 \times 2}{2\sin 27°} = 40.3\text{ N}$

## Example 2

In Fig. 1.1.12 there is no pivot but the principle of moments still applies – it is a universal rule. It could be a springboard with two supports, A and B and a (rather light) diver at the end.

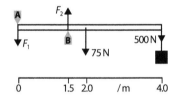

Fig. 1.1.12 Springboard to success

Where do we take moments about? We have a free choice – the PoM always applies. Let's make life easy: if we take moments about either **A** or **B** one of the forces plays no part and we'll just have an equation containing the other one. So,

### Answer

Taking moments about **A**: ΣACM = ΣCM,

$F_2 \times 1.5\text{ m} = 75\text{ N} \times 2.0\text{ m} + 500\text{ N} \times 4.0\text{ m}$

Leading to $F_2 = 1430\text{ N (3 s.f.)}$

Now take moments about **B**: ΣACM = ΣCM

$\therefore\ F_1 \times 1.5\text{ m} = 75\text{ N} \times 0.5\text{ m} + 500\text{ N} \times 2.5\text{ m}$

Leading to $F_1 = 860\text{ N (2 s.f.)}$

As the Pointer and Quickfire 12 show, there are more ways than this to solve the problem. In fact, taking moments about any point and then either resolving vertically or taking moments about any other point will give you enough information to do so.

# 1.1.5 Centre of gravity

The force of gravity acts on every particle in a body but, as far as solving problems is concerned, we can consider the weight of the body to be a single force acting through a point called the **centre of gravity**. For symmetrical bodies the C of G must lie on any plane of symmetry.

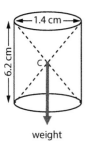 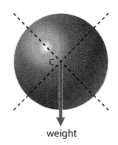 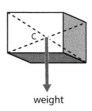

| weight | weight | weight |

Fig. 1.1.13 Centres of gravity

You'll use the concept of centre of gravity to answer questions such as the pub-sign and springboard in the previous section.

The position of the C of G is important for the stability of objects against toppling. The cylinder (or cuboid) in Fig 1.1.14 will be on the point of toppling if the C of G is just above the bottom left corner.

## Example

At what angle $\theta$ will the cylinder topple?

## Answer

The cylinder topples when the C of G is on the pecked line.

Using trig: $\tan \theta = \dfrac{1.4}{6.2}$, so $\theta = \tan^{-1} \dfrac{1.4}{6.2} = 12.7°$.

> **Key Term**
>
> The **centre of gravity** (C of G) of a body is the point where the entire weight of the body can be considered to act.

> **» Pointer**
>
> The pecked lines in Fig. 1.1.13 are not planes of symmetry but they do allow you to locate the centre of gravity.

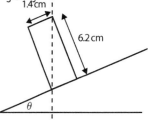

Fig. 1.1.14 Toppling cylinder

> **quickpire**
>
> ⑬ A cylinder, with a height of 15 cm and a radius of 0.5 cm, is standing on its end. Through how many degrees can it be tilted before it topples?

# 1.1.6 Specified practical work

## (a) Determining the density of a solid

This involves measuring the mass and volume and then using $\rho = \dfrac{m}{v}$.

The mass is measured using an electronic balance, with a typical resolution of 0.01 g or 0.1 g. The volume is determined by one of the following methods:

- Rules with mm scales, micrometers or digital callipers are used to measure the lengths and diameters of cuboids, cylinders and spheres and the volume calculated.

- Small irregular solids with a density greater than that of water are lowered into a measuring cylinder of water and the volume calculated from the change in water level reading.

- Larger irregular solids with a density greater than that of water are lowered into a full displacement can, the water which issues collected in a measuring cylinder and its volume measured.

> **quickpire**
>
> ⑭ A rectangular block of birch wood has dimensions in cm 6.52, 3.18 and 5.29. Its mass is 82.59 g.
>
> (a) Calculate its density.
>
> (b) The uncertainties in the length measurements are estimated as ±0.02 cm. Give the density with its absolute uncertainty.

## ≫ *Pointer*

The pivot in Fig 1.1.16 could be your outstretched finger. The pivot position is then only known to about ± 1 mm. With a knife edge and fine threads, ± 0.5 mm for each position is achievable.

### quicKfire

⑮ The C of G of a metre rule, with a mass of $(64.5 ± 0.1)$ g is at $(50.2 ± 0.1)$ cm on the rule's scale.

The rule balances with an unknown mass, $M$, at $(5.00 ± 0.05)$ cm and the pivot at $(32.6 ± 0.1)$ cm. Calculate:

(a) The values of $x$ and $y$ as in Fig. 1.1.16 together with their absolute uncertainties.

(b) The value of $M$ together with its absolute uncertainty.

## b) Determining masses using the principle of moments

This is done using a metre rule or, more conveniently, a ½-metre rule. The method compares the unknown mass with a known mass, e.g. standard 100 g masses or the mass of the rule.

With a **rule of unknown mass**, the C of G is found by locating the balance point of the rule without additional masses.

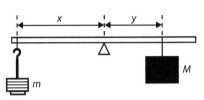

The ruler is balanced at the C of M, loaded as shown and $x$ and $y$ measured.

Fig. 1.1.15 Finding an unknown mass (1)

Taking moments about the pivot:
$mgx = Mgy, ∴ mx = My$.

With a **rule of known mass**, $m$, the C of G is found as above and the rule balanced with the unknown mass as shown. The lengths $x$ and $y$ are measured and the unknown mass, $M$, determined using the same equation as above.

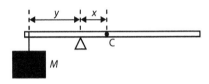

Fig. 1.1.16 Finding an unknown mass (2)

This experiment can be combined with volume measurements to find the density of the object of unknown mass.

## Extra questions

1. The speed, $v$, of tsunami waves is thought to be related to the ocean depth, $d$, by the equation $v = \sqrt{gd}$, where $g$ is the acceleration due to gravity. Show that the equation is homogeneous.

2. A website shows the speed of waves on water (of much smaller wavelength than tsunamis) is given by the equation: $c = \sqrt{\dfrac{g}{k} + \dfrac{\sigma k}{\rho}}$

   where $k = \dfrac{2\pi}{\lambda}$, is called the wave number, $\lambda$, $\gamma$ and $\rho$ have the usual meanings and $\sigma$ is a quantity called the *surface tension* of water.

   (a) Use the rules of homogeneity to express the unit of $\sigma$ in terms of the base SI units.

   (b) Show that the equation is dimensionally correct.

   (c) The unit of $\sigma$ is often given as N m$^{-1}$ or J m$^{-2}$. Show that these two units are equivalent with each other and with the expression you obtained in part (a).

3. The radius of the orbit of Venus is 0.723 AU, where the AU, *astronomical unit*, is the mean distance of the Earth from the Sun, $1.496 \times 10^{11}$ m. The orbital period is 224.7 days. Calculate the orbital speed of Venus.

4. (a) Find the sum of the following forces: 28 N due South and 45 N due West.

   (b) A car changes its velocity from 16.5 m s$^{-1}$ due North to 5.20 m s$^{-1}$ due West in 10 s. Calculate the change in velocity and the mean acceleration.

5. The sledge in Section 1.1.2(c) is subject to three additional forces: its weight, the snow exerts a vertically upwards force of 211 N and a frictional force of 400 N.

   (a) Show that the weight of the sled is 422 N.

   (b) Calculate the resultant force on the sledge. (Remember: magnitude and direction.)

   (c) Show that its acceleration is 1.23 m s$^{-2}$.

6. A drum of copper wire is sold as containing 1 kg of wire of diameter 0.193 mm. What length of wire would you expect it to contain? [$\rho_{Cu} = 8960$ kg m$^{-3}$]

7. A 25 g lump of modelling clay is placed at the 10 cm point of a uniform half-metre rule of mass 40 g. Find the balance point of the metre rule.

8. A uniform metre rule, of mass 80 g, is balanced on two knife edges placed at the 10 cm and 65 cm marks as shown.

   A block of metal of mass 100 g is placed on the metre rule.

   (a) Calculate the upward force on the rule by each knife edge when the block is placed: (i) at the left-hand end of the rule and (ii) at the 50 cm mark.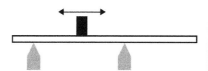

   (b) Where must the metal block be placed so that the upward force from the left-hand knife end is zero?

9. A micrometer is used to measure the diameter of the wire in question 6 in different places and taking readings at right angles. The following results were obtained:

   Diameter readings / mm: 0.191, 0.194, 0.192, 0.194, 0.195, 0.185, 0.194.

   (a) Discuss whether these results are consistent with the stated 0.193 mm diameter.

   (b) Determine the length of wire on the drum with its uncertainty [take mass = 1.000 kg].

> ## Pointer
>
> We use a line over a quantity, e.g. $\bar{v}$ to denote the mean value.

## quickpire

① Use the same method to calculate $\overline{v_{AB}}$. You'll need to use a bit of trig or Pythagoras' theorem.

## quickpire

② State the instantaneous velocity:
(a) at A
(b) 30 s from the start.

# 1.2 Kinematics

You will recognise a lot of this area of study from your pre-A level course. It aims to develop the GCSE concepts of motion and apply them in more advanced ways, such as motion in two dimensions.

## 1.2.1 Definitions

We'll use the example of the car on the test track to sort out a few definitions. The length of the track (i.e. the circumference) is 3.6 km and the car completes one lap at a constant speed in 80 s, starting at **A**.

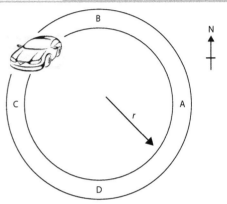

Fig 1.2.1 Displacement, speed and velocity

1. What is the **mean speed** (in one lap)?

   Using the equation in the Key terms:

   $$\text{Mean speed} = \frac{\text{total distance travelled}}{\text{total time taken}} = \frac{360 \text{ m}}{80 \text{ s}} = 45 \text{ m s}^{-1}.$$

   In this case, because the speed is constant, the mean speed is also equal to the **instantaneous speed** at every instant.

2. What is the **mean velocity** in the first 40 s, (i.e. between **A** and **C**)?

   We need to know the displacement. You should be able to show that the diameter of the track is 1150 m (3 s.f.) so the displacement from **A** to **C** is 1150 m West. The time is 40 s.

   ∴ Mean velocity **A→C** =

   $$\overline{v_{AC}} = \frac{\text{total displacement}}{\text{total time taken}} = \frac{1150 \text{ m West}}{40 \text{ s}} = 29 \text{ m s}^{-1} \text{ West.}$$

3. What is the **instantaneous velocity** at **D**?

   Because the speed is constant we know that the magnitude of the velocity at **D** is 45 m s$^{-1}$, so we just need to add the direction, which is due East.

   ∴ $v_D = 45$ m s$^{-1}$ East.

4. What is the **mean acceleration** in the first 20 s (between **A** and **B**)?

   The change in velocity, $\Delta v = v_B - v_A$. If you've forgotten how to subtract vectors, look back at Section 1.1.2(b). From the vector diagram, Fig. 1.2.2:
   $\Delta v = 45\sqrt{2} = 64$ m s$^{-1}$ (2 s.f.)

So the mean acceleration $\bar{a} = \dfrac{\Delta v}{\Delta t} = \dfrac{64}{20} = 3.2$ m s$^{-2}$. = 3.2 m s$^{-2}$.

5. What is the **mean acceleration over one complete lap**?

The initial velocity and the final velocity between **A** and **A** are both 45 m s$^{-1}$ North, so the change in velocity, $\Delta v = 0$.

∴ Mean acceleration = 0

Notice that the car is always accelerating even though it is going at a steady speed. Acceleration is the rate of change of *velocity*, not speed.

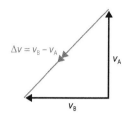

Fig. 1.2.2 Change in velocity

## 1.2.2 Graphical representation of motion

### (a) Displacement–time (x–t) graphs

The important thing to remember about $x$–$t$ graphs is that the gradient is the velocity. So

- Straight line = constant gradient = constant $v$
- Graph curving upwards = increasing gradient = increasing $v$, i.e. acceleration
- Negative gradient = negative, i.e. reverse, $v$.

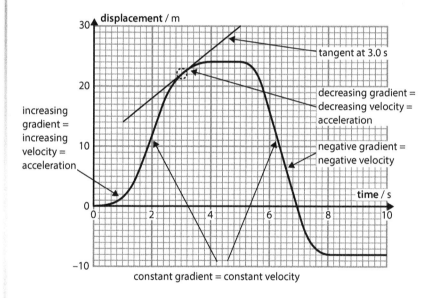

Fig. 1.2.3 Displacement-time graph

### Example

For Fig. 1.2.3, calculate the velocity at (a) 2.0 s, (b) 3.0 s

### Answer

(a) Between 1.6 s and 2.4 s the velocity is constant. The displacement increases from 6.5 m to 17.5 m.

$$\therefore v = \frac{\Delta x}{\Delta t} = \frac{17.5 \text{ m} - 6.5 \text{ m}}{2.4 \text{ s} - 1.6 \text{ s}} = \frac{11.0 \text{ m}}{0.8 \text{ s}} = 14 \text{ m s}^{-1} \text{ (2 s.f.)}.$$

### ≫ Pointer

The symbol $\Delta$ means 'change of', i.e. if a variable x changes its value from $x_1$ to $x_2$:

$$\Delta x = x_2 - x_1$$

∴ for example, $\bar{a} = \dfrac{\Delta v}{\Delta t}$

### quickfire

③ State the mean velocity over one complete lap.

### quickfire

④ In Fig. 1.2.3 the velocity is zero between 4.0 s and 5.0 s. When else is it zero?

### quickfire

⑤ A student's qualitative description of the motion shown in Fig. 1.2.3 starts:

Acceleration from rest forwards from 0 to 1.6 s, then constant velocity to 2.4 s....

Continue this description for the whole journey.

(b) First you need to draw the tangent to the graph at 3.0 s (see the graph). The velocity at 3.0 s = the gradient of the tangent.

$$\therefore v = \frac{\Delta v}{\Delta t} = \frac{30.0 \text{ m} - 14.0 \text{ m}}{5.0 \text{ s} - 1.0 \text{ s}} = \frac{16.0 \text{ m}}{4.0 \text{ s}} = 4.0 \text{ m s}^{-1}.$$

## ≫ *Pointer*

If the graph is below the line, the velocity is negative so the displacement is decreasing. The 'area' between the graph and the time axis is therefore considered to be negative.

## ≫ *Pointer*

The gradient of the tangent to a non-linear $v$–$t$ graph is the instantaneous value of the acceleration.

## Grade boost

The word *deceleration* is often used instead of *negative acceleration* when an object is slowing down, i.e. its speed is decreasing. This is a problem when the velocity is negative, e.g. 15 – 18 s on the graph: here the gradient is negative, the velocity is decreasing (becoming more negative) but the speed is increasing. Take care to say what you mean!

## quickᴘɪʀᴇ

⑥ Between what times does the $v$–$t$ graph show:
(a) Positive acceleration?
(b) Increasing speed?
(c) Constant velocity?
(d) Negative acceleration?
(e) Decreasing speed?

## (b) Velocity–time ($v$–$t$) graphs

This is the most common graph to appear in exam questions. There are two important points to remember:

1. The gradient of the line is the acceleration (because the gradient is the rate of change of velocity with respect to time).

2. The area between the line and the time axis gives the displacement.

The following is a typical graph beloved of examiners:

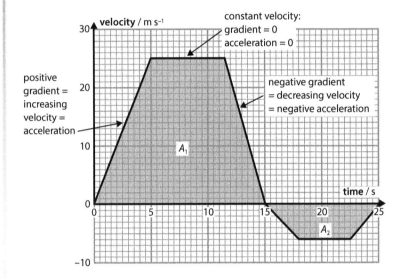

*Fig. 1.2.4 Velocity-time graph*

Again we'll have a couple of questions. There are some follow-up ones later.

**Example**

Calculate:

(a) The acceleration for the periods 0–5 s and 5–11.5 s.

(b) The mean velocity between 15 and 25 s.

**Answer**

(a) Acceleration 0 – 5 s = $\dfrac{\Delta v}{\Delta t} = \dfrac{(25 - 0) \text{ m s}^{-1}}{(5.0 - 0.0) \text{ s}} = 5.0 \text{ m s}^{-2}$

Acceleration 5 – 11.5 s = 0 (constant velocity)

(b) The mean velocity, $\bar{v} = \dfrac{\Delta x}{\Delta t}$, where $\Delta x$ is the displacement, which is the negative 'area' $A_2$. $A_2$ is the area of a trapezium:

$\Delta x = A_2 = \frac{1}{2}(10 + 5) \times 6 = -45$ m  (see Fig. 1.2.5)

$\therefore \ \bar{v} = \dfrac{\Delta x}{\Delta t} = \dfrac{-45 \text{ m}}{10 \text{ s}} = -4.5$ m s$^{-1}$

Note that, in an exam, there would be nothing wrong in working out $A_2$ as + 45 m or mean velocity as 4.5 m s$^{-1}$ and then adding the comment that it was in the negative (or reverse) direction.

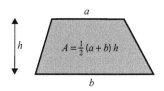

Fig. 1.2.5 Area of a trapezium

## (c) Acceleration–time (*a–t*) graphs

### Example

Without any calculations, sketch an acceleration–time graph for the motion shown in Fig. 1.2.4.

**Answer** (see Pointer)

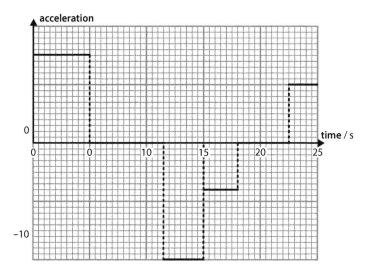

Fig. 1.2.6 Acceleration-time graph

>> *Pointer*
Acceleration–time graphs are not explicitly mentioned in the syllabus. However, there is no reason why an examiner should not ask you to draw one.

>> *Pointer*
The vertical pecked lines on the *a–t* grid indicate where the acceleration changes. The values of the acceleration have not been calculated.

### quicKpire

⑦ Calculate the values of acceleration shown in Fig. 1.2.4 and use them to redraw Fig. 1.2.6.

# 1.2.3 Constant acceleration kinematic equations

Remember that these equations are only valid for constant acceleration. You need to be able to derive as well as to use them.

## (a) Deriving the equations

The first one arises almost immediately from the definition of acceleration:

By definition: $a = \dfrac{v - u}{t}$ so, rearranging, $at = v - u, \therefore v = u + at.$ [1]

Notice that we are assuming that the time starts from 0, so $\Delta t = t$, the time taken.

Next, we need a $v$–$t$ graph for constant acceleration: Fig 1.2.7(a)

The displacement $x$ is the trapezium area

$\therefore$ Using the trapezium formula:

$$x = \tfrac{1}{2}(u + v)t \qquad [2]$$

If we divide the area as in Fig. 1.2.7(b) we can calculate the displacement differently:

$$x = A_1 + A_2$$

where $A_1$ is a rectangle and $A_2$ is a triangle:

$$A_1 = ut$$

$$A_2 = \tfrac{1}{2}\,\text{base} \times \text{height}$$

$$= \tfrac{1}{2}t \times (v - u)$$

But, from [1],

$$\therefore A_2 = \tfrac{1}{2}t \times at = \tfrac{1}{2}at^2$$

$$\therefore x = ut + \tfrac{1}{2}at^2 \qquad [3]$$

Deriving equation [4] requires more algebra:

Start from equation [1] and re-arrange it: $t = \dfrac{v - u}{a}$

Then substitute for t in one of the other equations: [2] is easiest and gives: $x = \tfrac{1}{2}\dfrac{(u + v)(v - u)}{a}$

Multiplying by $2a$ and multiplying out the brackets gives: $2ax = v^2 - u^2$

$\therefore$ Making $v^2$ the subject gives: $v^2 = u^2 + 2ax$ [4]

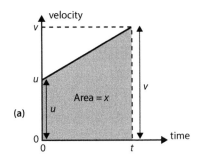

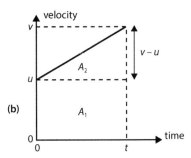

Fig. 1.2.7 Deriving equations of motion from a v-t graph

---

### quickfire

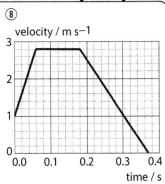

From the above $v$–$t$ graph:

(a) Draw an acceleration–time graph.

(b) Calculate the total displacement.

### ≫ Pointer

The derivation of
$$v^2 = u^2 + 2ax$$
makes use of the standard expansion
$$(a + b)(a - b) = a^2 - b^2.$$
If you don't understand why this is, try multiplying out the brackets yourself.

### ⚑ Grade boost

If you like algebra, have a go at deriving the equation:

$$x = vt - \tfrac{1}{2}at^2$$

It's not in the Data Booklet but it is sometimes useful.

## (b) Using the equations

The equations relate the variables $x$, $u$, $v$, $a$ and $t$. Notice that all four equations have four of these, i.e. each has one missing. The examiner will usually give you the values of three of these and ask you to calculate a fourth – use the equation which contains just these four variables.

For example: the examiner tells you that a car accelerates from rest ($u = 0$) up to a velocity ($v$) in a certain time ($t$) and asks you to calculate the displacement ($x$). You use the equation that contains $x$, $u$, $v$ and $t$,

i.e. $x = \frac{1}{2}(u + v)t$. You'd use the same equation if $x$ were given and $t$ was to be calculated.

### Examples

1. A formula 1 car decelerates uniformly from 100 m s⁻¹ to rest in 175 m. Calculate the time taken for the car to stop.

2. Calculate the displacement of a spacecraft, as it accelerates at 1.5 m s⁻² from a velocity of 15 km s⁻¹ to 18 km s⁻¹.

### Answers

1. Using $x = \frac{1}{2}(u + v)t$

   Make $t$ the subject → $t = \dfrac{2x}{u + v}$

   ∴ The time $t = \dfrac{2 \times 175 \text{ m}}{(100 + 0) \text{ m s}^{-1}} = 3.5$ s

   | | |
   |---|---|
   | $x = 175$ m | $v = 0$ |
   | $u = 100$ m s−1 | $t = ?$ |

2. Using $v^2 = u^2 + 2ax$

   Either: $18\,000^2 = 15\,000^2 + 2 \times 1.5x$

   ∴ $3.24 \times 10^8 = 2.25 \times 10^8 + 3x$

   ∴ $x = \dfrac{3.24 \times 10^8 - 2.25 \times 10^8}{3}$ m $= 3.3 \times 10^7$ m

   i.e. Displacement = 33 000 km

   | | |
   |---|---|
   | $x = ?$ | |
   | $a = 1.5$ m s⁻² | |
   | $u = 15$ km s⁻¹ | |
   | $v = 18$ km s⁻¹ | |

### Using the quadratic formula

Unless $u = 0$, using equation 3, $x = ut + \frac{1}{2}at^2$, to calculate $t$ requires the use of the quadratic formula. As usual, this formula gives two answers (or none!) and so you'll have to select the appropriate one.

### Example

How long to does it take a train with initial velocity 10 m s⁻¹ to travel 500 m if accelerating at 2.5 m s⁻²?

### Answer

Using $x = ut + \frac{1}{2}at^2$, $500 = 10t + 1.25t^2$

Rearranging: $1.25t^2 + 10t - 500 = 0$

∴ $t = \dfrac{-10 \pm \sqrt{100 + 2500}}{2.5} = -61$ s or 41 s – which is right? See Pointer.

| | |
|---|---|
| $x = 500$ m | $u = 10$ m s⁻¹ |
| $a = 1.5$ m s⁻¹ | $t = ?$ |

 **Pointer**

Examiners often state that an object accelerates 'from rest'. This means that the initial velocity, $u = 0$.

 **Grade boost**

When answering questions, make a list of the variables you know and which you need to calculate.

**Pointer**

It is your choice whether to swap round the algebraic equation (as in answer 1) or to put in the data first (answer 2).

 **Grade boost**

In Example 2, there is nothing wrong in working in km. In this case $a = 0.0015$ km s⁻², so
$18^2 = 15^2 + 2 \times 0.0015x$
which leads to

$x = \dfrac{324 - 225}{0.003}$ km $= 33\,000$ km.

 **Pointer**

In this case the time of −61 s clearly cannot be right, so the answer must be 41 s. Sometimes it is slightly trickier to decide. See the next section.

## Grade boost

Look out for the examiner's statement, 'ignoring air resistance' or 'air resistance is negligible.' In this case the downward acceleration is 9.81 m s$^{-1}$.

### Key Term

The **acceleration due to gravity** is the acceleration of a freely falling object, i.e. one on which gravity is the only force.

### quickfire

⑨ A netball is dropped from rest from a height of 15 m. Calculate its speed when it hits the ground. (Note: the algebra gives two answers. Comment on which is correct.)

## Grade boost

At the highest point, the cricket ball has velocity = 0. To calculate this height put $v = 0$ and use $v^2 = u^2 = 2ax$. You should be able to show that the maximum height is 22.7 m.

### quickfire

⑩ For the cricket ball:

(a) Calculate the time at which the cricket ball is 15.0 m above the ground. Why are there two answers?

(b) Why can't you find the time at which the ball is 25 m above the ground?

# 1.2.4 Falling vertically under gravity

## (a) Without air resistance

In these questions (if they relate to the Earth), you won't be told the acceleration in the question as it will always be the **acceleration due to gravity** (or the acceleration of *free fall*) which the Data Booklet gives as 9.81 m s$^{-2}$. The constant acceleration equations in Section 1.2.3 can be applied.

Questions often include upwards as well as downwards motion, so we need to be careful about directions. When answering a question we define either upwards or downwards (it doesn't matter which) as the positive direction.

If upwards is positive:

- The acceleration $= -g = -9.81$ m s$^{-2}$.
- Downward velocities are negative.
- Positions above the initial level have a positive displacement; points below have a negative displacement.

When the direction is always downwards (see Quickfire 9) just take downwards as positive, without comment.

### Example

A cricket ball is hit vertically upwards at 22 m s$^{-1}$. Calculate its height after 2.0 s.

### Answer

Taking upwards as positive:

Using $x = ut + \frac{1}{2}at^2$, $x = 22 \times 2 - \frac{1}{2} \times 9.81 \times 2^2$.

$\therefore x = 24.4$ m

So the height is 24.4 m.

$$x = ?$$
$$g = -9.81 \text{ m s}^{-1}$$
$$u = 22 \text{ m s}^{-1}$$
$$t = 2.0 \text{ s}$$

## (b) With air resistance – terminal velocity

Air resistance (drag) depends upon the velocity through the air – the faster you go, the greater the drag. This means that the resultant force and the acceleration are not constant, so we cannot use the equations of motion from Section 1.2.3. The questions which you will be asked are not, in principle, more difficult than those at GCSE but the examiner will require greater clarity in your answers!

The examiner's favourite standby is the free-fall parachutist or skydiver. Fig 1.2.8 gives a typical $v$–$t$ graph, with annotations:

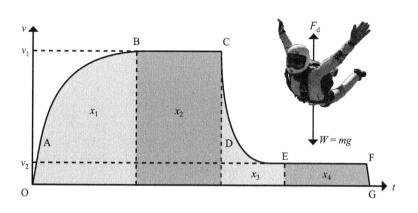

Fig. 1.2.8 v–t graph for a skydiver

Make sure you can answer the questions in the Grade boost (see Extra questions).

## Grade boost

Typical questions:
Explain the features of the graph [long QER question].
What is the gradient at O?
Why does the gradient decrease from A to B?
What do the areas $x_1 - x_4$ represent?
At what point is the parachute deployed?
Why are there two different terminal velocities, $v_1$ and $v_2$?
What feature of the graph represents the total descent distance; the deceleration at the end?

# 1.2.5 Projectiles – ballistic motion

The black circles in Fig. 1.2.9 represent the motion of an object which is projected horizontally and then moves freely (i.e. no thrust and no air resistance) under gravity.

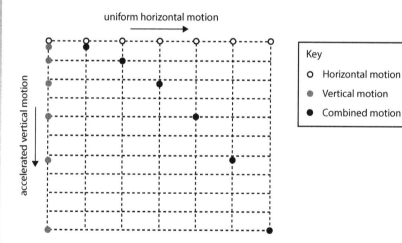

uniform horizontal motion

accelerated vertical motion

Key

o   Horizontal motion
●   Vertical motion
●   Combined motion

Fig. 1.2.9 Independence of horizontal and vertical motion

The diagram shows that we can treat the motion as being composed of two independent motions:

- Uniform horizontal velocity.
- Uniform vertical acceleration ($= g$).

The easiest projectile questions are those for which the angle of projection is zero. The following example illustrates how we treat the horizontal and vertical motions separately:

## Pointer

Fig 1.2.9 represents snapshots of the position of the object at equal time intervals. The black dots move horizontally to the right at a constant speed and move downwards at a constant acceleration.

## quickfire

⑪ If the time interval between the positions of the dots is 0.20 s, calculate the distance between the horizontal lines.

Hint: Calculate the distance fallen by an object, in (say) 1.2 s.

## >> Pointer

Quite often people use $y$ for the vertical displacement when dealing with projectiles and $x$ for the horizontal displacement. It is also common to write $g$ for the acceleration (or $-g$ if upwards is positive).

## quickfire

⑫ Use both methods to work out the velocity with which the rock hits the sea.

## Example

A rock is thrown horizontally at 30 m s⁻¹ from the top of a 50 m high sea cliff. Calculate where it enters the sea.

### Answer

**Vertical motion**: Initial velocity = 0. Take downwards as positive

To calculate the time taken to fall 50 m:

Using $y = u_y t + \frac{1}{2} g t^2$ (see Pointer)

Then $50 = 0 + \frac{1}{2} \times 9.81 t^2 \therefore t = \sqrt{\dfrac{2 \times 50}{9.81}} = 3.193$ s

$$u_y = 0$$
$$y = 50 \text{ m}$$
$$a = 9.81 \text{ m s}^{-2}$$
$$t = ?$$

**Horizontal motion**: Constant velocity of 30 m s⁻¹

$\therefore$ Distance from cliff = speed × time = 30 m s⁻¹ × 3.193 s = 96 m (2 s.f.)

If the question asks for the velocity with which the rock hits the water then you could:

- Use the calculated time (from above) and $v_y = u_y + gt$ to work out the vertical velocity, **or**
- Use $v^2 = u^2 + 2gy$ to calculate the vertical velocity, **then**
- Combine $v_x$ and $v_y$ to find the resultant velocity (remembering to calculate the direction) (see Quickfire 10).

## A standard AS question

This question is deliberately written to be quite difficult. Often the examiner will break it down into bite-size chunks!

## >> Pointer

If you're nervous of using the quadratic formula, try this approach:

1. Find $u_x$ and $u_y$ as in the example.
2. Find the vertical velocity when it hits the ground using
   $v^2 = u^2 + 2ax$
   ($v_y^2 = u_y^2 - 2gy$)
   for $y = -1.99$ m, remembering to choose the negative root.
3. Find the time for hitting the ground using:
   $v_y = u_y - gt$.

### Example

A shot putter launches the shot from a height of 1.99 m as shown. Calculate the horizontal distance travelled by the shot before it lands.

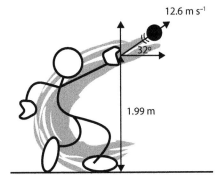

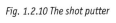
*Fig. 1.2.10 The shot putter*

### Answer

The first thing we need to do is to resolve the initial velocity, $u$, into its horizontal and vertical components $u_x$ and $u_y$. Then we can treat the horizontal and vertical motion separately as in the previous example.

1. Taking upwards as positive
   $u_x = 12.6 \cos 32° = 10.69$ m s⁻¹; $u_y = 12.6 \sin 32° = 6.68$ m s⁻¹.

2. **Vertical motion**:

   The shot hits the ground when $y = -1.99$ m

   Using $y = u_y t - \frac{1}{2}gt^2$, $-1.99 = 6.68t - 4.905t^2$

   Rearranging: $4.905t^2 - 6.68t - 1.99 = 0$

   $$\therefore t = \frac{6.68 \pm \sqrt{(-6.68)^2 - 4 \times 4.905 \times (-1.99)}}{9.81} = 1.614 \text{ s or } -0.252 \text{ s}$$

   Ignoring the negative root → time to hit the ground = 1.614 s

3. **Horizontal** range = $u_x t = 10.69$ m s$^{-1}$ × 1.614 s = 17.3 m

$$a = -g$$
$$y = -1.99 \text{ m}$$
$$u_y = 6.68 \text{ m s}^{-1}$$
$$t = ?$$

**quickpire**

⑬ Use the method in the Pointer to solve the shot putter problem.

# 1.2.6 Specified practical work – measuring $g$ by free fall

The easiest technique is to drop a small heavy object from a tall building and measure the time it takes to reach the ground – see Fig. 1.2.11. Galileo showed that, in the absence of air resistance, all small heavy objects have the same acceleration, so the actual mass is irrelevant.

Measurements:

1. The vertical height $h$, using a tape measure or a laser ranger.
2. The drop time, $t$, using a stopwatch. To improve accuracy a signalling system is needed, e.g. using flags.

The analysis:

Using: $x = ut + \frac{1}{2}at^2$, with $x = h$, $a = g$ and $u = 0$

$$\therefore h = \frac{1}{2}gt^2 \rightarrow g = \frac{2h}{t^2}.$$

Fig 1.2.11 A new use for the leading tower of Pisa

The problem with this technique is that even for a 20 m high building, the drop time ~2 s. A 3 s drop time needs about 45 m. So we turn to electronics to help us out:

Fig. 1.2.12 shows apparatus to record the descent time of a steel ball bearing. The switch to turn off the electromagnet also starts the timer. The timer is stopped when the ball hits the flap. The timer records the descent time to ± 0.01 s. If the ball is dropped from a series of different heights up to ~70 cm (say. 20, 30 ......70) and a graph drawn of $h$ against $t^2$ the gradient is $g/2$ (so g is 2 × the gradient).

**Note:** The electromagnet typically takes a few centi-seconds to release the ball. You should be able to show that, in this case the graph of $h$ against $t^2$ will not be a straight line, but a graph of $\sqrt{h}$ against $t$ will be – with a gradient of $\sqrt{\frac{1}{2}g}$ and a positive intercept on the $t$ axis.

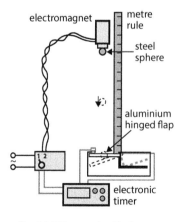

Fig. 1.2.12 Improved method

# Extra questions

1. The velocity–time graph is for a car of mass 800 kg driving between two sets of traffic lights.

   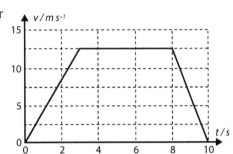

   Determine:

   (a) The initial acceleration.

   (b) The total distance travelled.

   (c) The mean velocity.

   (d) The maximum resultant force on the car.

2. For the displacement–time graph in Fig. 1.2.3

   (a) Identify the time period for which the $x$–$t$ graph shows the maximum speed.

   (b) Calculate the maximum speed shown.

   (c) Calculate the mean acceleration between 2.0 s and 6.0 s.

3. For the velocity–time graph in Fig 1.2.4:

   (a) Identify the times at which the velocity is zero.

   (b) Determine the total time for which (i) the speed and (ii) the velocity is less than 10 m s$^{-1}$.

   (b) Calculate the mean velocity for the whole journey.

4. For the skydiver's $v$–$t$ graph in Fig. 1.2.8:

   (a) Identify the significance of the areas $x_1 - x_4$.

   (b) State the gradient of the $v$–$t$ graph in Fig. 1.2.8 at O.

   (c) Explain the shape of the graph between C and G.

   (d) The gradient of the velocity–time graph for the 80 kg skydiver at a certain instant between A and B is 4.5 m s$^{-2}$. Determine the drag force on the skydiver at this instant.

5. A cricket ball is struck at 24.7 m s$^{-1}$ at an angle of 45° to the horizontal. Assuming that air resistance can be ignored and that the initial height of the ball is negligible, calculate:

   (a) The initial horizontal and vertical components of velocity.

   (b) The greatest height reached by the cricket ball.

   (c) The horizontal distance the ball travels before it hits the ground.

6. If the cricket ball in question 5 strikes the vertical wall of a building 3.0 s after being hit, calculate:

   (a) The velocity with which it hits the wall.

   (b) The height above the ground at which it hits the wall.

# 1.3 Dynamics

This part of A level Physics is based upon Newton's laws of motion, which deal with the action of forces on objects. They may be summarised as:

1. A body's velocity will be constant unless a force acts upon it.

2. The rate of change of **momentum** of a body is directly proportional to resultant force acting upon it.

3. If a body **A** exerts a force on body **B**, then **B** exerts an equal and opposite force on **A**.

These laws are often referred to as N1, N2 and N3 for brevity.

**» Pointer**

A car approaching a bend on an icy road could well fall victim to N1. Without the frictional grip of the tyres it would plough straight on!

## 1.3.1 The concept of force

Newton's 1st law (N1) gives the rather surprising statement that objects carry on moving in a straight line with a constant speed unless something makes them change this behaviour. This is an idealisation which is outside our everyday experience because we always have friction, air resistance or other dissipative forces to slow us down.

The air track is a piece of apparatus where the rider floats on a cushion of air. This reduces friction to a very low level and the rider only very gradually slows down.

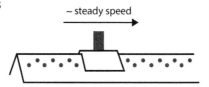
~ steady speed

*Fig. 1.3.1 Air track*

**Key Term**

A **force** is that agency which changes (or tends to change)* the velocity of a body.

* Often more than one force acts. If they are in equilibrium then there will be no change of velocity.

So, if we see an object speeding up, slowing down or changing direction we say, 'Aha! There's a **force** at work here'. To anticipate N2, the direction of the force is the direction of the velocity change. We should also be able to identify the object which is exerting the force.

**Grade boost**

Sometimes students find it difficult to identify the pairs of forces in N3 but it is easier than you might think. Just swap the objects around and say, 'equal and opposite', e.g. a **tractor** pulls a **trailer** – so its pair force is 'a **trailer** pulls a **tractor** with an equal and opposite force'.

### Test yourself

Identify the source of the force on X and its direction in the following diagrams. The arrows show the direction of motion of X.

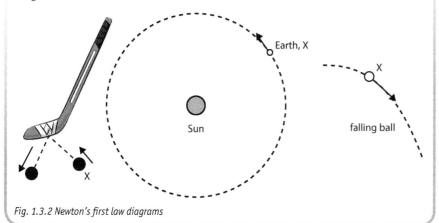

*Fig. 1.3.2 Newton's first law diagrams*

## quicKpire

① Describe the N3 'equal and opposite forces' to those shown in the diagram.

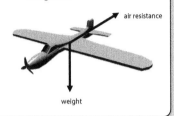

air resistance

weight

## quicKpire

② Identify the forces acting on the skydiver and the N3 equal and opposite forces:

## >> Pointer

Avoid referring to the normal contact force as the normal reaction. This sounds like a poor statement of N3.

## quicKpire

③ In QF2, assuming that the skydiver is at terminal velocity, which forces are equal and opposite but are not N3 pairs? Explain.

## quicKpire

④ Identify the N3 pair forces in the Earth/Sun system in Fig. 1.3.2.

# 1.3.2 Newton's 3rd law of motion

In all the situations above, there are two objects involved; stick and ball; Sun and Earth; ball and Earth. This is always the case: the hockey stick exerts a force on the ball ($F_1$) and the ball exerts a force ($F_2$) on the hockey stick. And N3 tells us that these forces are equal and opposite.

i.e. $F_2 = -F_1$.

In case you don't believe that this force, $F_2$, exists, just look at the dents in this used hockey stick.

If you'd like another quick demo of N3, just kick a brick wall. Your foot will exert a force on the wall (it might even cause a bit of damage) and the wall will exert a force on your foot (hence the pain).

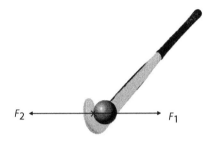

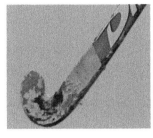

*Fig. 1.3.3 The effect of N3 on a hockey stick!*

### Example

A box sits on a table. Give the N3 equal and opposite force to each of the named forces (weight and normal contact force) in Fig. 1.3.4.

### Answer

The weight is the gravitational force that the Earth exerts on the box. So the 'equal and opposite' force to this is the

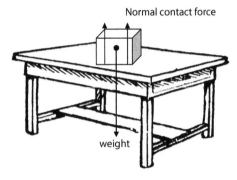

*Fig. 1.3.4 N3 forces on a box*

gravitational force that the box exerts on the Earth, i.e if the box weighs 10 N, it pulls the earth upwards with a force of 10 N!

The normal contact force is the upwards force exerted by the table on the box, so the equal and opposite force to this is the downwards force exerted by the box (in fact these are both electromagnetic forces arising because the atoms in the box and table surfaces are being slightly squashed).

**Warning**: Some forces are equal and opposite, but they are not Newton's 3rd law pairs. So **don't fall into the following trap:**

The weight is equal and opposite to the normal contact force so they are a pair of N3 equal and opposite forces.

This might seem sensible at first glance but it is wrong! There are three rules that must be obeyed:

1. The forces act on different bodies [but the weight and contact force both act on the same body].
2. The forces are equal and in opposite directions.
3. The forces are of the same type, e.g. both gravitational [this is clearly not the case for the normal contact force and the weight].

## 1.3.3 Free-body diagrams

Consider the (slightly artificial) system of a table and box Fig. 1.3.5(a). If we want to identify the interactions between the box and the other objects shown, several of the forces will be on top of one another. **Free-body diagrams**, in which the relevant objects are separated, help us around this.

Here the box and the table have been separated so the forces which act on each are clearly separated. This way of doing it helps in identifying the N3 pairs.

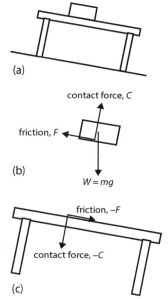

- The contact force, $C$, on the box is clearly applied by the table, so there must be an equal and opposite force ($-C$) on the table. It doesn't matter which of the pair you think of first – they arise simultaneously.
- The box is held from slipping down by the frictional force, $F$, applied by the table, so there must be an equal and opposite force on the table ($-F$).

Fig. 1.3.5 Free-body diagrams

- The weight ($W$) of the box is the gravitational force of the Earth on the box; the N3 pair is not shown here as it acts on the Earth as a whole.

### Example

Fig. 1.3.6(a) shows a skydiver at the final stage of the descent. Identify the significant forces on the parachute and the skydiver by drawing free-body diagrams. Identify relationships between the forces assuming terminal velocity has been reached.

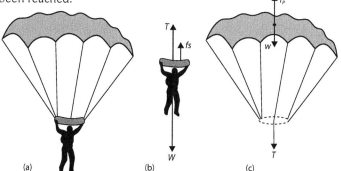

Fig 1.3.6 Forces on a parachute and parachutist

>> **Pointer**
Avoid the expression 'action and reaction' for the N3 pair. There is no first force and second force – they occur at the same time. The supposed statement of N3, 'Action and reaction are equal and opposite' is inadequate and may attract no credit in an exam.

## quickpire

⑤ N3 is *always* valid – not just when objects are in equilibrium.

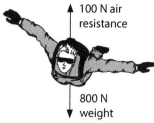

100 N air resistance
800 N weight

The forces acting on the skydiver are as shown. Identify the N3 partners, the objects they act on, their magnitudes and directions.

On the skydiver – Fig. 1.3.6(b):

$W$ = weight of skydiver; $f_S$ = air resistance on skydiver;

$T$ = supporting force provided by tension in parachute chords.

$$W = T + f_S \qquad (1)$$

On the parachute – Fig. 1.3.6(c):

$w$ = weight of chute; $T$ = force exerted by skydiver on parachute;

$F_P$ = air resistance on parachute.

$$T + w = F_P \qquad (2)$$

From (1) and (2) $W + w = F_P + f_S$.

## 1.3.4 Force and momentum

**quickfire**

⑥ Show that N s is equivalent to kg m s⁻¹.

The linear momentum, $p$, of a body is defined by

$$p = mv.$$

The unit of momentum is therefore $kg\ m\ s^{-1}$. An alternative unit is N s (the reason for this is explained below).

Newton's 1st and 2nd laws (N1 and N2) tell us that a force produces a proportional rate of change of momentum in a body. So, if a body's momentum is changing, the resultant force, $F$, on it is given by

$$F = k\frac{\Delta p}{\Delta t} = k\frac{\Delta (mv)}{\Delta t},$$

**quickfire**

⑦ Calculate the momentum of:

(a) A car of mass 1000 kg travelling at 20 m s⁻¹.

(b) A proton of mass $1.66 \times 10^{-27}$ kg travelling at $3.00 \times 10^7$ m s⁻¹.

where $k$ is a constant. In our unit system, SI, we define the unit of force, the newton (N), so that the value of $k$ is 1 exactly. Hence, in SI,

$$F = \frac{\Delta p}{\Delta t} = \frac{\Delta (mv)}{\Delta t}.$$

This relationship explains the use of N s as the unit of momentum. $\Delta p = F\Delta t$, so the unit of $p$ (the same as the unit of $\Delta p$) is the product of the unit of $F$ and the unit of $t$.

### Example

A gun sprays a jet of water of cross-sectional area 5 mm² [$= 5 \times 10^{-6}$ m²] at a speed of 30 m s⁻¹. Calculate the force which the gun exerts on the water. [$\rho_{water}$ = 1000 kg m⁻³]

*Fig. 1.3.7 Force on water jet*

**» Pointer**

Force is change of momentum per second.

**quickfire**

⑧ In the Example, identify the N3 partner to the force on the water. Sketch a free-force diagram for the gun.

### Answer

Consider a time of 1 s:

Volume, $V$, of water sent out = $5 \times 10^{-6} \times 30 = 1.5 \times 10^{-4}$ m³.

∴ mass of water sent out = $\rho V = 1.5 \times 10^{-4} \times 1000 = 0.15$ kg

The water starts out with 0 momentum.

∴ Change of momentum of water in 1 s = $0.15 \times 30 = 4.5$ N s.

∴ Force exerted on water = momentum change per second = 4.5 N

## Momentum–time graphs

From its definition, the gradient of a momentum–time graph is the resultant force on a body. For example, this graph shows the variation of momentum with time for a car travelling in a fixed direction over a period of 6 seconds.

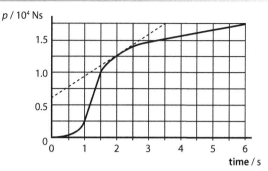

Fig. 1.3.8 Momentum–time graph

The gradient of the tangent at 2 s (pecked line) is the resultant force on the car at this time. [See Quickfire 9]

**quickfire**

⑨ For the $p$–$t$ graph, in Fig. 1.3.8, determine:

(a) The resultant force at 2.0 s.

(b) The resultant force between 1.0 s and 1.5 s.

(c) The mean force between 0 and 6.0 s.

## 1.3.5 Force and acceleration

If there are several forces acting on a single object, it is useful to rewrite the above equation as

$$\Sigma F = \frac{\Delta (mv)}{\Delta t},$$

where the $\Sigma$ sign (the Greek letter capital sigma) indicates 'the sum of', so $\Sigma F$ means the sum of the forces. For an object with a constant mass, $m$, we can re-write this equation as:

$$\Sigma F = m\frac{\Delta v}{\Delta t},$$

The quantity $\frac{\Delta v}{\Delta t}$ is the acceleration, $a$, so the equation becomes:

$$\Sigma F = ma \quad \text{or} \quad a = \frac{\Sigma F}{m}$$

This equation tells us that the acceleration of a body is the resultant force divided by the mass. The resultant force is the sum of all the forces acting on the body. [Remember that forces are vectors, so the 'sum' is the vector sum.] If the forces balance, then $\Sigma F = 0$, the body is in equilibrium and the acceleration is zero. If the forces are not balanced, then $\Sigma F \neq 0$ leading to an acceleration. You really need to know how to use this equation!

### Example

As an easy example, calculate the acceleration of the skydiver in Quickfire 5.

### Answer

Resultant force $\Sigma F = 800 - 100 = 700$ N (downwards).

We need to calculate the mass. The gravitational field strength of the Earth is 9.81 N kg$^{-1}$. This gives us a mass of 81.5 kg.

$\therefore$ Acceleration, $a = \frac{\Sigma F}{m} = \frac{700}{81.5} = 8.6$ m s$^{-2}$.

**quickfire**

⑩ A ball of mass 0.057 kg is thrown at a wall. Its velocities just before and just after hitting are 17 m s$^{-1}$ east and 13 m s$^{-1}$ west respectively. Find:

(a) Its change in momentum.

(b) The mean force on the ball if the ball and wall are in contact for 0.015 s.

[Remember that $\Delta p$ and $F$ are both vectors, so a direction is needed!]

**quickfire**

⑪ What mean force does the ball exert on the wall in QF 10?

**≫ Pointer**

The value for $g$, the Earth's gravitational field strength (9.81 N kg$^{-1}$) is numerically the same as the acceleration due to gravity (9.81 m s$^{-2}$). These are given in the Physics Data Booklet, which you'll have in your exams.

## » Pointer

The direction of the acceleration is the same as that of the resultant force.

### quickfire

⑫ Calculate the acceleration of the block.

14.2 N ← [ 2.8 kg ] → 9.7 N

### quickfire

⑬ Calculate the acceleration of the golf ball.

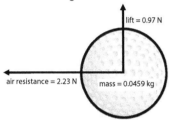

lift = 0.97 N

air resistance = 2.23 N    mass = 0.0459 kg

## » Pointer

Momentum can be conserved even if there are forces from outside a system, providing these forces have zero resultant force, e.g. the vertical forces on a rider on an air track.

---

If there are two or more forces on an object in different directions we have to use the vector law of addition to find the resultant force.

### Example

Calculate the acceleration of the box in Fig. 1.3.9.

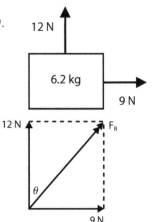

### Answer

Step 1: Calculate the resultant force, $F_R$, by drawing the parallelogram of vectors and using Pythagoras' theorem.

$$F_R^2 = 12^2 + 9^2 = 144 + 81 = 225$$
$$\therefore F_R = \sqrt{225} = 15 \text{ N}$$

Step 2: Calculate $\theta$. Why? Because acceleration is a vector and examiners expect you to give a magnitude and a direction. You could equally well calculate the angle between $F_R$ and the 9 N force.

$$\tan \theta = \frac{\text{opp}}{\text{adj}} = \frac{9 \text{ N}}{12 \text{ N}} = 0.75 \therefore \theta = 37° \text{ (2 s.f.)}$$

Fig. 1.3.9 Calculating acceleration using resultant force

Step 3: Calculate $a$: $a = \dfrac{\Sigma F}{m} = \dfrac{15}{6.0} = 2.5$ m s$^{-2}$, 37° from the 12 N force.

# 1.3.6 The principle (or law) of conservation of momentum

This law states: *The vector sum of the momenta of the bodies in an isolated system is constant.*

This is very condensed language. This is what it means:

If we study a system which is a set of objects, we measure the momentum of each and then add these values together (remembering they are vectors) we will get the same answer at some later time as long as: (i) no particles have entered or left the system and (ii) there are no forces acting on the particles from outside the system [this is what 'isolated' means in this context].

It might seem that this law can't be very useful because there are always forces acting from outside any system. It is useful in studying collisions because the forces from outside, e.g. friction, are often very small compared with the collision forces between the bodies themselves. The law has been confirmed in a vast number of sub-atomic particle collisions.

## (a) Deriving conservation of momentum from N3 and N2

Consider two bodies **A** and **B**, e.g. two positive ions, which repel each other, with forces $F_A$ and $F_B$.

From N3. $F_A = -F_B$

Fig. 1.3.10 Particle interactions

Then, assuming there are no other forces from outside the system:

$$\frac{\Delta p_A}{\Delta t} = \frac{\Delta p_B}{\Delta t} \qquad \text{so} \qquad \Delta p_A = -\Delta p_B$$

In other words, A's change in momentum over any interval of time is equal and opposite to B's, so the *vector sum* of the momenta never changes. This applies to a system of any number of bodies. Also the forces $F_A$ and $F_B$ don't have to be repulsive – $F_A$ will always be equal and opposite to $F_A$.

The simplest examples involve two objects, one of which is initially stationary, which collide and then combine to produce a single object.[1]

### Applying the law: a simple example

A 5000 kg wagon travelling at 9 m s$^{-1}$ collides with a stationary wagon of mass 1000 kg. If they couple on impact, calculate their velocity after the collision.

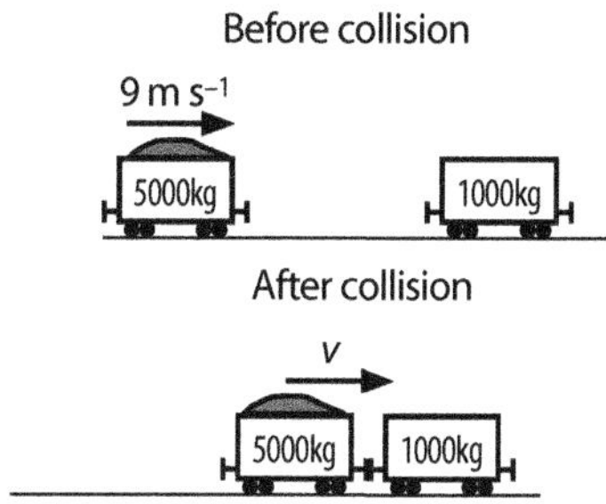

### Answer

We'll write out the answer in a lot of detail.

Before collision

9 m s$^{-1}$

5000kg        1000kg

After collision

$v$

5000kg  1000kg

Fig. 1.3.11 Momentum conservation 1

1st step:  Draw a diagram (if none provided). Label known and unknown quantities.

2nd step:  Write the C of M equation:

Vector sum of initial momenta  =  Vector sum of final momenta

3rd step:  Write in the *mv* products

$$5000 \times 9 + 1000 \times 0 = 6000v$$

Simplify ∴ $\qquad\qquad 45\,000 = 6000v$

Rearrange ∴ $v = \dfrac{45000}{6000} = 7.5$ m s$^{-1}$

4th step:  Write your answer:

∴ Common velocity after collision = 7.5 m s$^{-1}$ to the right

That example was simple because all the motion was in the same direction. Velocity and momentum are vectors so, if there is motion in opposing directions we need to define a positive direction. Vectors in that direction are positive; those in the other direction are negative.

---

[1]  Historically, Galileo made the first experiments towards the law using this approach.

## >> Pointer

Note the handling of momentum as a vector:
- The initial momentum of the 5 kg sphere 9.0 N s to the left, which is −9.0 N s to the right.
- The question asked for the *velocity* so you must give the direction.

The next example has different directions and also the objects bounce off each other.

### Applying the law: a more complicated example

The diagrams contain data about a pair of spheres which collide. Find the velocity, $v$, of the 2.0 kg sphere after the collision.

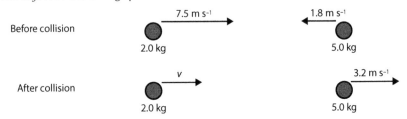

*Fig 1.3.12 Momentum conservation 2*

### Answer

Omitting units and considering momentum to the right [i.e. right is +ve]:

Vector sum of initial momenta = Vector sum of final momenta

So $\qquad 2.0 \times 7.5 - 5.0 \times 1.8 = 2.0v + 5.0 \times 3.2$

So $\qquad 6.0 = 2.0v + 16$

So $\qquad v = -5.0 \text{ m s}^{-1}$.  [Note '−' sign]

∴ Velocity of the 2.0 kg sphere = 5.0 m s⁻¹ to the left

## quickfire

⑯ Calculate the energy lost in the 'spheres' example.

## quickfire

⑰ A favourite example of bellicose examiners is the recoil of a gun: A 1200 kg field gun fires a 20 kg shell horizontally at a speed of 300 m s⁻¹. Assuming external forces can be ignored, calculate:
(a) The recoil velocity of the gun.
(b) The total kinetic energy.

## >> Pointer

Most everyday collisions are somewhere between elastic and perfectly inelastic, i.e the objects don't stick together on impact but some kinetic energy is converted to internal energy in the objects.

## (b) Elastic and inelastic collisions

Just like momentum, the **total** energy of an isolated system is conserved but, as we know from the principle of conservation of energy, it can be converted from one form to another. Because of this the **kinetic energy** of a system is **not always conserved**. Looking at the last example, during the actual contact both balls slow down temporarily, so their total kinetic energy must be less at this time – they regain kinetic energy as they rebound.

An **elastic collision** *is one in which no kinetic energy is lost or gained* (once the bodies have separated).

At room temperature, collisions between atoms of noble gases (such as helium) are elastic. But objects made of many atoms – that includes all objects we can see – collide **inelastically**: some of their KE is transferred to additional random vibrational energy of their atoms. The maximum loss in KE occurs if the objects stick together on impact (like the wagons in the example): such collisions are sometimes called **pefectly inelastic**.

### Example

Calculate the energy lost in the wagon collision in Section 1.3.6(a)

### Answer

Initial kinetic energy = $\frac{1}{2} \times 5000 \times 9^2 = 202\ 500$ J

Final kinetic energy = $\frac{1}{2} \times 6000 \times 7.5^2 = 168\ 750$ J

∴ Energy lost = 202 500 − 168 750 = 34 000 J (2 s.f.)

# 1.3.7 Specified practical work – investigating Newton's 2nd law

Taking N2 in its form $F = ma$, investigating the law involves two experiments:

1. Showing that $a \propto F$ for a constant $M$
2. Showing that $a \propto \dfrac{1}{M}$ for a constant $F$

In school, these investigations are normally carried out using an air track.

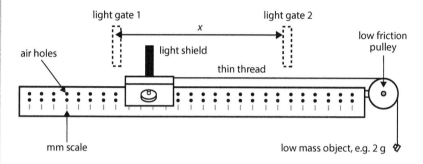

Fig. 1.3.13 Air track used for $F = ma$

The accelerating force is $mg$ where $m$ is the mass of the object on the end of the thread.

Both investigations require the velocity of the rider to be determined as it passes a light gate. This is done by dividing the width of the light shield by the time for which the light is cut off (measured using a digital timer). Both also require the acceleration to be determined.

- **Method 1:** Using two light gates (as shown in Fig. 1.3.13).
  The rider is released to the left of light gate 1 and the velocities measured (as above) as it passes light gates 1 and 2 $u$ and $v$). The displacement, $x$ is measured using the mm scale and the acceleration calculated using $v^2 = u^2 + 2ax$.

- **Method 2:** Using one light gate.
  The rider is released from rest ($u = 0$) , the time, $t$, for it to accelerate through displacement $x$ measured using a stopwatch and its velocity, $v$, at that point measured as above. The acceleration is calculated using $v = u + at$.

## ≫ *Pointer*

The small mass, $m$, is also accelerated.

In **investigation 1**, the overall mass can be kept accurately constant by initially loading up the rider with all the small masses which it is intended to use. Each small mass is transferred to the end of the thread as required.

In **investigation 2**, the graph which is plotted should be a $a$ against $\dfrac{1}{M+m}$.

### (a) Investigation 1: Showing that $a \propto F$ for a constant $M$

The accelerating force is varied by using a series of different masses on the end of the thread. A graph is plotted of $a$ against $m$ and a straight line through the origin is obtained. As the accelerating force is proportional to $m$, this confirms the relationship (but see Pointer).

### (b) Investigation 2: Showing that $a \propto \dfrac{1}{M}$ for a constant $F$

The acceleration measured for a series of different masses $M$ (obtained by loading the rider) a graph of $a$ against $\dfrac{1}{M}$ plotted and a straight line through the origin obtained. This confirms the relationship.

## Extra questions

1. A sliding ice-hockey puck gradually decelerates in its horizontal motion; its vertical motion is not accelerated.

   (a) Draw a free-body diagram for the puck, identifying the forces on the puck
   (b) Discuss them in terms of Newton's three laws of motion.

2. A fully-laden railway truck of total mass 40 tonnes collides at 5.0 m s$^{-1}$ with a stationary truck of mass 5.0 tonnes. The two trucks couple automatically on impact. Assuming external forces can be ignored calculate:

   (a) Their common velocity immediately after impact.
   (b) The percentage loss in kinetic energy.

3. In a repeat of the collision in question 2, the trucks fail to couple. A trackside observer reports that the unladen truck has a speed of 10.0 m s$^{-1}$ immediately after the collision. Use the principles of conservation of momentum and energy to show that this is cannot be true.

4. A free-fall sky-diver of mass 80 kg (including parachute, etc.) has an acceleration of 6.0 m s$^{-2}$ when her downward velocity is 40 m s$^{-1}$. [$g = 9.81$ m s$^{-2}$]

   (a) Calculate the magnitude of the air resistance on her.
   (b) Assuming that the air resistance is proportional to the square of her velocity, calculate:
   (i) Her acceleration when she is falling at 50 m s$^{-1}$.
   (ii) Her terminal velocity.

5. [More difficult] A garden hose has a right-angled bend in it. Water flows along the pipe as shown.

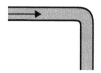

(a) Explain why the water in the pipe suffers a change in momentum as it goes around the bend and find its direction.

(b) Explain in terms of N2 and N3 why the hose experiences a force and state its direction.

(c) [Much more difficult!] The hose has an internal diameter of 1.5 cm and the rate of flow of water is 250 cm³ s⁻¹. Calculate the force exerted on the hose.

# 1.4 Energy concepts

## 1.4.1 Work

If a force is applied to an object which moves, the **work** done by the force on the object is calculated using the formula

$$W = Fx \cos \theta,$$

where    $F$ is the force

        $x$ is the distance moved

        $\theta$ is the angle between the force and the direction of movement

so      $x \cos \theta$ is the distance moved in the direction of the force.

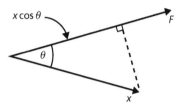

Fig. 1.4.1 Definition of work

>> **Pointer**

$F \cos \theta$ is the component of $F$ in the direction of motion, so the work can also be thought of as the product of the distance moved and the component of the force in the direction of motion.

Often (but not always) in AS exams the force and motion are in the same direction, i.e. $\theta = 0$, and we know that cos 0 = 1, so the equation simplifies to

$$W = Fx$$

When a force does work it can speed things up, lift them higher, make things hotter, etc. In other words it causes an energy transfer. In fact we define energy in such a way that the **energy transfer is equal to the work done**.

So if a force does 100 J of work on an object, then 100 J of energy is transferred to the object. Notice that the units of work and energy are the same: joule (J).

### quickfire

① How far does the engine move before the force has done 1 GJ of work?

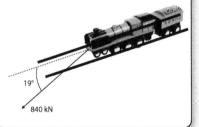

19°

840 kN

In the case of the engine in Quickfire 1, we know how much energy is transferred (1 GJ) but we don't know in what form the energy is. Because of friction and air resistance it won't all be kinetic.

**Example**

A woman pulls a sled along a flat snow field for 0.90 km. Calculate the work done by the woman.

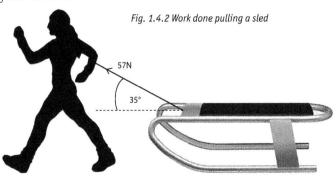

Fig. 1.4.2 Work done pulling a sled

57N

35°

>> **Pointer**

Make sure that your calculator is in the correct mode when using trig functions. In this case it needs to be in DEG not RAD.

## Answer

This is just a straightforward 'put the numbers into a formula' question but you have to remember to convert the km to m (0.90 km = 900 m):

$$W = Fx \cos \theta$$
$$= 57 \times 900 \times \cos 35$$
$$= 42 \text{ kJ (2 s.f.)}$$

Notice that the work done does not depend on how quickly the woman pulls the sled. The work done is 42 kJ whether she takes 10 minutes or 30 minutes to complete the journey. The shorter the journey the greater the woman's **power** (see Section 1.4.5)

### Some cases to look out for

$\theta = 0$ is a special case, here are some more:

1. $\theta = 90°$ [$\frac{\pi}{2}$ rad].
   In this case $\cos \theta = 0$, $\therefore$ $W = 0$.

2. If $\theta > 90°$
   In this case $\cos \theta < 0$, $\therefore$ $W < 0$.

3. If $\theta = 180°$ [$\pi$ rad]
   In this case $\cos \theta = 0$, $\therefore$ $W = -Fx$

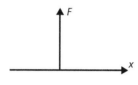
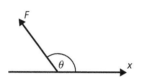
Fig. 1.4.3 Forces in different directions

An example of the last case is an object sliding to a halt. The frictional force on the object is in the opposite direction to the motion, so the angle between $F$ and $x$ is 180° and the force does a negative amount of work. We'll come back to this in the context of energy.

# 1.4.2 Energy formulae

We can use the idea that

Work done = energy transfer

to derive the formulae for various forms of energy.

## (a) Kinetic energy

Imagine a stationary object of mass $m$ which is acted on by a single constant force, $F$. It accelerates in the direction of the force. When it has moved a distance, $x$, its speed is $v$.

Fig. 1.4.4 Kinetic energy derivation

quicKɸɪɾe

③ Calculate the KE of a 20 tonne lorry travelling at 30 m s$^{-1}$.

quicKɸɪɾe

④ A helium atom has a kinetic energy of $8.3 \times 10^{-22}$ J. Calculate its speed. (mass = $6.6 \times 10^{-27}$kg)

 **Pointer**

The **change** in kinetic energy, $\Delta E$ of an object is given by

$$\Delta E = \tfrac{1}{2}mv^2 - \tfrac{1}{2}mu^2.$$

We can factorise this to give

$$\tfrac{1}{2}m(v^2 - u^2)$$

**WARNING:** $\Delta E \neq \tfrac{1}{2}m(v - u)^2$.

This is a common student mistake.

quicKɸɪɾe

⑤ A constant force of 0.85 N accelerates a toy car from 1.2 m s$^{-1}$ to 5.5 m s$^{-1}$ over a distance of 13.2 cm. Calculate the mass of the car.

The work done by the force is given by $W = Fx$ [because $\theta = 0$].

We know that the acceleration a is given by $F = ma$ from Newton's second law of motion.

The 4th of the equations of motion is: $v^2 = u^2 + 2ax$

But $u = 0$ in this case, so $\qquad\qquad v^2 = 2ax$

Multiplying by $\tfrac{1}{2}m$ $\qquad\qquad \tfrac{1}{2}mv^2 = max$

But $F = ma$ (Newton's 2nd law), $\therefore$ $\quad \tfrac{1}{2}mv^2 = Fx$

In that last equation, $Fx$ is the work done by the force [because $\theta = 0$], so the energy transfer to the object must be $\tfrac{1}{2}mv^2$. Because it didn't have any kinetic energy to start with [it wasn't moving!], this must be the formula for kinetic energy.

**Example**

A car of mass 800 kg, travelling at 15 m s$^{-1}$ has a constant resultant accelerating force of 500 N. How far must it travel to reach a speed of 40 m s$^{-1}$.

**Answer**

We could solve this using one of the equations of motion but we are going to use work and energy.

The change in kinetic energy, $\Delta E = \tfrac{1}{2}mv^2 - \tfrac{1}{2}mu^2 = \tfrac{1}{2}m(v^2 - u^2)$

$$= \tfrac{1}{2} \times 800(40^2 - 15^2)$$

$$= 550\,000 \text{ J}$$

Change in kinetic energy = work done by the resultant force

$$\therefore Fx = 500x = 550\,000$$

$$\therefore x = \frac{550\,000}{500} = 1100 \text{ m}$$

So the car reaches a speed of 40 m s$^{-1}$ after travelling 1.1 km

## (b) Gravitational potential energy

We'll approach this in the same way as for kinetic energy, but this time our thought experiment involves lifting a body of mass $m$ at constant velocity (so it doesn't gain any kinetic energy) in the absence of air resistance (so the air molecules do not gain any kinetic energy).

The force, $F$, required to lift the body is equal and opposite to the gravitational force, $mg$, on the body.

If the body is raised through a height $\Delta h$, the work done is given by

$$W = F\Delta h = mg\Delta h.$$

The work done is equal to the energy transfer. The only gain in energy is gravitational potential energy.

Hence: the increase in gravitational potential energy, $\Delta E = mg\Delta h$.

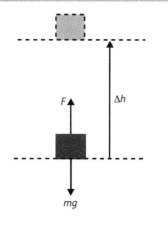

Fig. 1.4.5 Gravitational potential energy derivation

There are a few things to notice about gravitational potential energy:

1. Unlike kinetic energy, there is no obvious place where potential energy is zero. It is only changes in potential energy which are significant.

2. Potential energy can be negative: if, for convenience, we take the potential energy to be zero at ground level, anything below ground level has negative potential energy.

**》 Pointer**

When raising an object in this way, the work is done *by* the force, $F$, *against* the force of gravity.

**▲ Grade boost**

Grade boost
The formula $\Delta E = mg\Delta h$ assumes that $g$ is constant. This is true as long as $\Delta h$ is much less than the Earth's radius (6370 km). The year 13 course has a formula for calculating $\Delta E$ in an inverse square law gravitational field.

## (c) Elastic potential energy

Springs and stretched rubber bands and bent rulers can do work on things in returning to their normal shape (e.g. by firing a projectile). We cannot calculate the work done in stretching an object just by multiplying the force by the extension because the force is not constant – it increases with the stretch. Instead we use the area under the force–extension ($F$–$x$) graph.

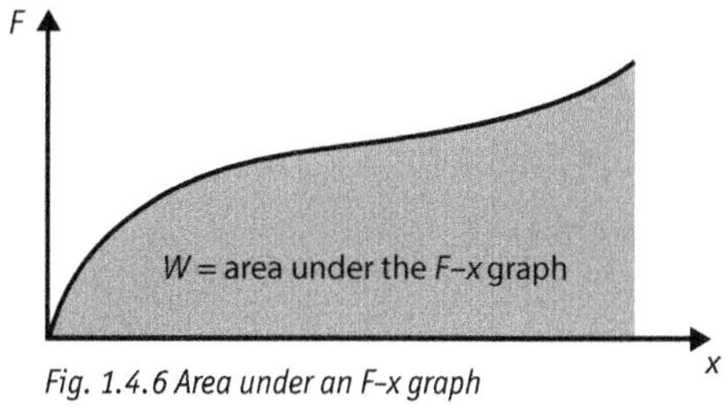

In the case of a stretched spring which is operating in its elastic region the force and extension are proportional and related by

$$F = kx$$

where $k$ is the **spring constant**.

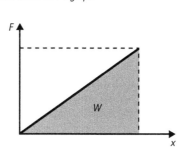

Fig. 1.4.6 Area under an F–x graph

$W$ = area under the $F$–$x$ graph

Fig. 1.4.7 Work done stretching a spring

**quickᖴire**

⑥ Calculate the GPE lost by a 75 kg skydiver in a 10 km jump.

**Key Term**

**Elastic potential energy** is the energy possessed by an object by virtue of its deformation

(or, in everyday language, ....because of it has been stretched, squashed or bent).

## Grade boost

The area under a graph of $y$ against $x$ is physically significant if the product $y\Delta x$, for constant $y$, has a meaning. $F\Delta x$ is the work done by a constant force, so, in the language of calculus, $W = \int F \, dx$, which is the area under the $F$–$x$ graph.

We can calculate $W$ from the formula for the area of a triangle, $A = \frac{1}{2}bh$,

so the work done in stretching the spring (or other material which obeys Hooke's law) is given by

$$W = \tfrac{1}{2} Fx$$

Combining this with $F = kx$ gives two more formulae

$$W = \tfrac{1}{2} kx^2 \text{ and } W = \tfrac{1}{2} \frac{F^2}{k}$$

A 100% elastic material can do the same amount of work when it is allowed to return to its normal size or shape, so the same formulae give the elastic potential energy, i.e.

$$\text{Elastic potential energy, } E_p = \tfrac{1}{2}kx^2 = Fx = \tfrac{1}{2} \frac{F^2}{k}$$

These same formulae apply to any elastic object with a linear force–extension curve. Catapults and bows generally have non-linear $F$–$x$ curves, so we need to use the area under the $F$–$x$ curve to calculate the energy stored.

### quickpire

⑦ A spring has a constant, $k$, of 87 N m$^{-1}$. Calculate the extension and the elastic potential energy when it is extended by a force of 3.1 N.

### Symbols

The Data Booklet gives $W$ and $E$ as the symbols for work and energy respectively. Many books also use $W$ for energy and others use $E_k$ and $E_p$ for kinetic and potential energy (elastic and gravitational).

### Key Term

**The principle of conservation of energy** states that energy cannot be created or destroyed only transferred.

# 1.4.3 The principle of conservation of energy

If the block in Fig. 1.4.8 falls freely from rest through a height, $h$, it is easy to show, by using $v^2 = u^2 = 2ax$, that it attains a velocity, $v$ given by

$$v^2 = 2gh$$

Multiplying by $\tfrac{1}{2} m$ we have: $\tfrac{1}{2} mv^2 = mgh$

Now  $mgh$ is the *loss* in gravitational potential energy and

  $\tfrac{1}{2} mv^2$ is the *gain* in the kinetic energy

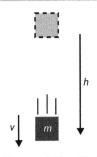

Fig. 1.4.8 Falling object

This is an illustration of the **principle of conservation of energy**:

- The KE + PE are the same.
- Energy is transferred from gravitational potential to kinetic.

Sometimes the energy stays associated with one body, as in the case of the falling block. In other cases, such as a collision, energy can be transferred from one object to another. Study the following slightly tricky example:

### Example

A spring has a stiffness 40 N m$^{-1}$. It is held vertically in a clamp stand. A load of mass 0.40 kg is attached to the unstretched spring and released. Calculate its speed when it has fallen a distance of 10 cm.

**Answer**

Let the initial gravitational potential energy be 0

When the load has fallen through 10 cm [= 0.10 m]

Gravitational potential energy = $mg\Delta h = 0.4 \times 9.81 \times (-0.10) = -0.392$ J

Elastic potential energy = $\frac{1}{2}kx^2 = \frac{1}{2} 40 \times 0.10^2 = 0.20$ J

Energy is conserved, so the total energy at all times = 0

∴ At 10 cm: KE + (−0.392 J) + 0.20 J = 0

∴ KE = 0.192 J

So the speed at 10 cm = 0.98 m s⁻¹ (see Quickfire 8)

# 1.4.4 The work–energy relationship

As we have seen, energy is defined together with work: the work done is equal to the energy transfer. An important example of this relationship, consider a force, $F$, applied to a body of mass m which has a speed, $u$. The diagram shows the case with the initial velocity and the force in the same direction:

Fig. 1.4.9 Work-energy relationship

Assuming that no other forces act (or that F is the resultant force), the equation leads to

$$Fx = \frac{1}{2}mv^2 - \frac{1}{2}mu^2$$

which is the mathematical expression of the statement that the work done on the body is equal to the change in kinetic energy. The relationship is more general than its derivation implies:
it applies also for $F$, $u$ and $v$ in any arbitrary directions. The object in the diagram below is initially moving with velocity $u$ and experiences a constant force, $F$, to the right. It follows the curved path and achieves a velocity of $v$ when the displacement is $x$, measured in the same direction as $F$.

The work done on the object is $Fx$ so, in the absence of other forces, the change of kinetic energy is given by

$$\frac{1}{2}mv^2 - \frac{1}{2}mu^2 = Fx$$

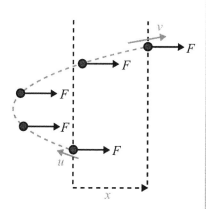

Fig. 1.4.10 Energy transfer when direction changes

> ## ≫ Pointer
> For the equation,
> $$Fx = \frac{1}{2}mv^2 - \frac{1}{2}mu^2$$
> to apply generally, the directions of the force and the displacement must be the same. Otherwise, we must use $Fx \cos \theta$.
> In this case both $F$ and $x$ are horizontal to the right.

### Example

The total mass of a cyclist and cycle is 92 kg. They slow down from 18 m s⁻¹ to 13 m s⁻¹ in a distance of 47 m. Calculate the mean force acting.

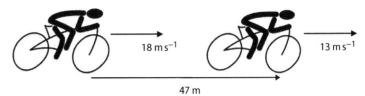

18 m s⁻¹

13 m s⁻¹

47 m

*Fig. 1.4.11 Decelerating cyclist*

### Answer

The force, $F$, displacement, $x$, and the initial and final velocities, $u$ and $v$, are related by:

$$Fx = \tfrac{1}{2}mv^2 - \tfrac{1}{2}mu^2$$

$$\therefore \quad F = \frac{\tfrac{1}{2}mv^2 - \tfrac{1}{2}mu^2}{x}$$

$$= \frac{\tfrac{1}{2}92 \times 13^2 - \tfrac{1}{2}92 \times 18^2}{48} \quad \text{(see Grade boost)}$$

$$= \frac{7774 - 14904}{47} = -150 \text{ N (2 s.f.)}$$

The minus sign indicates that the force is in the opposite direction to the displacement (as expected).

**Grade boost**

It is important to get the initial and final velocity the right way round. In this case the final velocity is less than the initial velocity so the change in kinetic energy is negative, so the work done is negative. The examiner is likely to notice if you get the wrong sign!

**Key Terms**

**Power** is the rate of energy transfer. If the energy transfer is due to work it can also be defined as the rate of doing work.

**Unit of power** = watt (W)

## 1.4.5 Power and energy

The work done by the woman in pulling the sled in the example in Section 1.4.1 is independent of the time she takes. However, her **power** is inversely proportional to the time she takes. The SI unit of power is the watt (W) which is equivalent to the joule per second, i.e. W = J s⁻¹.

### Example

The woman in the example takes 8 minutes 20 seconds to complete the journey. Calculate her power output.

### Answer

$$\text{The power } P = \frac{W}{t} = \frac{42\,000 \text{ J}}{500 \text{ s}} = 84 \text{ W}$$

### A note on units

The SI units of work (energy) and power are rather small for many everyday applications. A typical domestic kettle has a power of 2–3 kW. In 5 minutes a 2.5 kW kettle transfers 750 kJ, i.e. 0.75 MJ. A typical thermal power station has an electrical power output of 2 GW, giving an annual energy transfer to

**quickfire**

⑨ Criticise the following statement: 'In 2012 the total electricity generated was 38.4 GW per year.'

the national grid of 60 PJ. This use of SI multipliers is one way of handling the large numbers. Various other methods are commonly used:

- kW h. In the equation $\Delta E = P\Delta t$, if $P$ is expressed in kW and $\Delta t$ in hours, the energy transfer is expressed in kilowatt hours (kW h). MW h, GW h and TW h can also be used.

- tce (tonnes of coal equivalent): the energy released by burning 1 tonne of coal ~ 29.4 GJ. This quantity of energy is known as 1 tce. For international energy use, Mtce (= million tce) and Gtce.

# 1.4.6 Dissipative forces

When a force does work on an object against gravity or elastic tension or in producing acceleration, the object possesses energy (potential or kinetic). This is not the case when doing work against friction or air resistance. These two forces are known as **dissipative forces**.

For example, the force that the woman applies on the sled in Section 1.4.1 produces no acceleration because the horizontal component is exactly equal and opposite to the frictional force of the snow on the sled runner. And if the woman stops pulling, the kinetic energy of the sled will soon be lost. The work against dissipative force of friction results in energy being transferred to the random motion of the individual particles [molecules, electrons, ions...] of the runners and the snow. This is called **internal energy**.

The essential feature of internal energy is its random nature. The motion in kinetic energy is ordered. Potential energy requires all the particles of the system to have been displaced in an orderly way. This allows for 100% conversion between kinetic and potential energies. However, there is no way of coaxing all the randomly vibrating molecules in an object into moving in the same direction: the energy is there but it is much less useful.

Here are a couple of examples:

1. A bearing rotates against frictional forces. Working against the friction causes a wasteful increase in the internal energy the bearings [which is later transferred away by conduction, convection and radiation] and a reduction in the useful transfer of energy.

2. [The examiner's favourite] A skydiver falling through the air experiences drag. This results in a transfer of energy to eddy motions in the air which dissipate into internal energy of the air molecules. At terminal speed, the rate of loss of gravitational potential energy is equal to the rate of increase in atmospheric internal energy.

## Grade boost

Most exam questions involving practical energy transfers require the use of SI multipliers.

## quickfire

⑩ Express in TW h the annual energy transfer of a 2 GW power station.

## Key Terms

The **mechanical energy** of a system is the sum of its kinetic and potential energies.

**Dissipative forces** are forces which lead to a loss in mechanical energy in a system. Examples are dynamic friction, fluid resistance (e.g. air resistance or drag).

**Internal energy** is the sum of the random kinetic and potential energies of the individual particles in a system.

In everyday language this is often referred to as **thermal energy**. Never call it heat!

## quickfire

⑪ Explain in terms of forces and energy why a submarine providing a constant thrust will not keep on accelerating.

## Grade boost

'Analyse' is a tricky word. Here it means give an energy transfer equation with all the values calculated.

## Pointer

Remember that 1 tonne is $10^3$ kg.

### quickpire

⑫ Calculate the maximum power output of the train motors in the example.

### Key Term

**Efficiency** is the fraction of the input energy which is usefully transferred.

## Grade boost

It is useful to keep the units in the calculation.
$W\ m^{-2} \times m^2 = W$

### Example – with tricky wording!

A 4-car underground railway set has a mass of 40 tonnes. It accelerates from rest to 20 m s$^{-1}$ over a distance of 500 m. The driving force is a constant 18 kN. Analyse the energy transfer.

### Answer

1. Work input by motor, $W = Fx = 18\ kN \times 500\ m = 9.0\ MJ$

2. Gain in KE by the 4-car set,
$$E_k = \tfrac{1}{2}mv^2 - \tfrac{1}{2}mu^2$$
$$= \tfrac{1}{2} \times 40 \times 10^3 \times 20^2$$
$$= 8.0 \times 10^6\ J = 8.0\ MJ$$

3. ∴ Work done against dissipative forces $= 9.0\ MJ - 8.0\ MJ$
$$= 1.0\ MJ$$

This manifests itself as internal energy in the air, wheels, rails, etc.

| Work Input 9.0 MJ | → | Gain in kinetic energy 8.0 MJ | + | Gain in kinetic internal energy 1.0 MJ |
|---|---|---|---|---|

## 1.4.7 Efficiency

In most energy transfers, some of the energy is wasted. In the underground example in Section 1.4.6 it ended up as low grade internal energy. In fact that is the usual end for wasted energy. The fraction of the input energy which is usefully transferred is called the **efficiency** of the transfer. It is often expressed as a percentage and calculated using the equation:

$$\text{Efficiency} = \frac{\text{useful energy transfer}}{\text{total energy input}} \times 100\%.$$

This is the equation on the Data Booklet. If we divide the energy transfers by the time taken we get another useful form of the efficiency equation:

$$\text{Efficiency} = \frac{\text{useful power transfer}}{\text{total power input}} \times 100\%$$

The concept of efficiency applies generally – not just to mechanical energy transfer, as the next example illustrates:

### Example

A solar panel on a satellite has a collecting area of 1.51 m$^2$. It is orientated at right angles to solar radiation of intensity 1.36 kW m$^{-2}$. It delivers an output of 10.2 A at 40 V. Calculate its efficiency.

### Answer

Power of incident radiation $= 1360\ W\ m^{-2} \times 1.51\ m^2 = 2054\ W.$

useful output power $= VI = 40\ V \times 10.2\ A = 408\ W$

$$\text{Efficiency} = \frac{\text{useful power transfer}}{\text{total power input}} \times 100\% = \frac{408}{2054} \times 100\% = 19.9\%$$

# Extra questions

1. The Highway Code figure for the braking distance for a car travelling at 70 mph is 75 m.
   (a) Using the approximation that 1 mile = 1600 m, convert 70 mph to m s⁻¹.
   (b) Without calculation, state what energy transfer occurs during braking.
   (c) Estimate the maximum braking force for an 800 kg car.

2. A 1 kg pebble is thrown with a kinetic energy of 50 J from the top of a 40 m high sea cliff. Ignoring air resistance, calculate:
   (a) Its kinetic energy when it hits the sea.
   (b) Its speed of projection.
   (c) Its speed when it hits the sea.

3. A 44 tonne lorry brakes from 25.5 m s⁻¹ to rest in a distance of 58.8 m.
   (a) Analyse this process in terms of the conservation of energy.
   (b) Explain why the efficiency of the braking process is 0%.

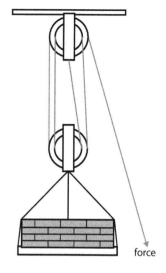
force

4. Some builders use a hoist to lift a load of bricks of mass 60 kg through a height of 15 m. To do so they pull the rope through a distance of 60 m with a force of 400 N. Calculate:
   (a) The increase in potential energy of the bricks.
   (b) The efficiency of the energy transfer.
   (c) Taking the masses of the lower pulley block and the pallet to be 5 kg and 2 kg respectively, analyse this process in terms of the conservation of energy.

5. The $F-x$ graph of a Tudor longbow is approximately as shown. It shoots a 55 g arrow with an efficiency of 88%. Estimate the speed of the arrow when the bow was drawn to its full distance of 70 cm before release.

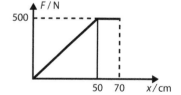

## ≫ Pointer

The equation $F = kx$ is not given in the Data Booklet. You need to learn it and Hooke's law.

## quickfire

① A spring extends by 10.0 cm when a load of mass 200 g is suspended from it. Calculate the spring constant, $k$

## quickfire

② Two springs, with spring constants 24.0 N m$^{-1}$ and 27.3 N m$^{-1}$ are joined end to end. Calculate the total extension of the pair when a load of mass 300 g is suspended from them.

# 1.5 Solids under stress

This topic deals with the way that different kinds of solid materials – metals, glasses, rubber – respond to being placed under tension. It starts with recapping the properties of springs.

# 1.5.1 Hooke's law and the spring constant

A spiral spring extends if it is under **tension**, i.e. it gets longer. The increase in length is called the **extension**.

Extension = stretched length – original length.

The arrangement shown in Fig. 1.5.1 can be used to investigate the relationship between the tension and the extension. Masses, $m$, are added and the scale readings indicated by the pointer noted. The extensions are calculated as above and the tension, $F$, is calculated using $T = mg$.

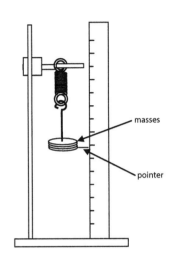

The graph of tension against extension is typically as shown in Fig. 1.5.2.

*Fig. 1.5.1 investigating a spring*

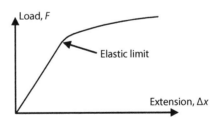

*Fig. 1.5.2 Tension-extension graph for a spring*

In the low-tension, straight-line region, the spring is **elastic**. For higher values of extension, above the **elastic limit**, the spring is permanently stretched. **Hooke's law** summarises this.

The relationship, $F = kx$, is valid for the elastic region. The constant, $k$, is called the spring constant. It is the force per unit extension of the spring. It is sometimes called the *stiffness* of the spring (or the *stiffness constant*). Many solid objects other than springs obey Hooke's law up to the elastic limit and the ratio of tension to extension for these objects is also often referred to as the stiffness constant.

**Example**

The spring constant $k$ for a spring is 25.3 N m⁻¹. The largest mass it can support without permanent extension is 0.65 kg. What is the extension at the elastic limit?

**Answer**

The tension, $F$, at the elastic limit $= mg = 0.65 \times 9.81 = 6.38$ N.

Using $F = kx$, the extension, $x = \dfrac{F}{k} = \dfrac{6.38 \text{ N}}{25.3 \text{ N m}^{-1}} = 0.25$ m

# 1.5.2 Stress, strain and the Young modulus

## (a) Definitions of $\varepsilon$, $\sigma$ and E

The cylinder (or wire) in Fig. 1.5.3 has a cross-sectional area $A$ and an original length $l_0$. It is under tension, $F$.

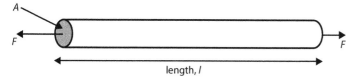

Fig. 1.5.3 Cylinder under tension

For a given load, the extension is proportional to the original length. So when comparing materials, engineers use the **strain**, $\varepsilon$, which is defined by

$$\varepsilon = \frac{\Delta l}{l_0},$$ where $\Delta l$ is the increase in length.

A similar sample of material with twice the cross-sectional area needs twice the tension to produce the same strain, so when comparing materials engineers work with the **stress**, $\sigma$, defined by

$$\sigma = \frac{F}{A}.$$

Hooke's law applies to most materials as well as springs. That means that $\Delta l \propto F$ for small extensions.

So, for any particular material, the strain is directly proportional to the stress and we define the **Young modulus**, $E$ by

$$E = \frac{\sigma}{\varepsilon}.$$

The value of $E$ is characteristic of the material: the values of $E$ enable us to compare the stiffness of materials – not the stiffness of particular specimens.

**Key Terms**

**Stress**, $\sigma$ = the tension per unit cross-sectional area.

Unit: pascal (Pa) or N m⁻².

**Strain**, $\varepsilon$ = the extension per unit length due to applied stress.

Unit: None

**Young modulus**, $E$ = the ratio of stress to strain for a material in the Hooke's law region.

Unit: pascal (Pa) or N m⁻²

**》 Pointer**

The specification and Data Booklet use the symbol $x$ for the extension in the formulae

$$W = \tfrac{1}{2} Fx \text{ and } E = \tfrac{1}{2} kx^2$$

They use $\Delta l$ in the formula

$$\varepsilon = \frac{\Delta l}{l_0}$$

**quickₚιre**

③ Express the unit of the Young modulus in terms of base SI units.

## » Pointer

A useful equation which is **not** in the Data Booklet is

$$E = \frac{Fl_0}{A\Delta l}$$

It can easily be derived from the three Data Booklet equations.

---

### quicKfire

④ Derive the formula

$E = \frac{Fl_0}{A\Delta l}$ from

$\sigma = \frac{F}{A}$, $\varepsilon = \frac{\Delta l}{l_0}$, and $E = \frac{\sigma}{\varepsilon}$

---

### quicKfire

⑤ A piece of wire has a strain of 0.001. What is the extension if the original length is:
(a) 1.0 m?
(b) 1.0 km?
(c) 53 cm?

---

### quicKfire

⑥ A load of 5 kN is applied to a steel rope of csa 2 cm². Calculate the stress in:
(a) N cm⁻²
(b) Pa
(c) MPa.

---

### Grade boost

With hard engineering materials (steel, concrete, glass) the value of the Young modulus is typically 10–200 GPa. If your answer is a lot smaller, you've probably not converted your units, e.g. cm to m².

## (b) Units of $\varepsilon$, $\sigma$ and E

From its definition, $e$ has no unit – it is one length divided by another. In most engineering situations, the values of $\varepsilon$ are very small, e.g. the girder of a bridge might be about 5 m long and extend by 0.5 mm when under tension.

In this example, $\varepsilon = \dfrac{5 \times 10^{-4}\ \text{m}}{5\ \text{m}} = 1 \times 10^{-4}$. The figure tells us that the girder extends by $1 \times 10^{-4}$ m for every metre of its length ($1 \times 10^{-4}$ cm for every cm…)

Looking at the definition of **stress**:

Unit of stress $= \dfrac{\text{unit of force}}{\text{unit of area}} = \dfrac{\text{N}}{\text{m}^2} = \text{N m}^{-2} = \text{Pa (pascal)}$. This is the same as the unit of pressure.

From the definition of the **Young modulus**

Unit of the Young modulus $= \dfrac{\text{unit of stress}}{\text{unit of strain}} = \dfrac{\text{Pa}}{1} = \text{Pa}$

## (c) Magnitudes of $\varepsilon$, $\sigma$ and E

The value of the strain we calculated in part (b) was very small – $1 \times 10^{-4}$. This is typical of strains which we encounter in most engineering materials which are behaving elastically. It is useful to bear this in mind when doing calculations, but strains in ductile metals and rubber can be much larger.

Values of stress tend to be large in most engineering materials. The girder in part (b) might be under a tension of 200 kN (the weight of a load of mass ~20 tonnes) and have a cross-sectional area of 100 cm², so the stress would be:

$$\sigma = \frac{200 \times 10^3\ \text{N}}{10^{-2}\ \text{m}^2} = 2 \times 10^7\ \text{Pa}\ (= 20\ \text{MPa}).$$

The values of stress in engineering situations vary greatly: MPa and GPa values are typical.

Because strain is typically $< 10^{-3}$, the values of the Young modulus for wood and metals are typically in the 10–500 GPa range. For the steel girder:

$$E = \frac{\sigma}{\varepsilon} = \frac{2 \times 10^7\ \text{Pa}}{1 \times 10^{-4}} = 2 \times 10^{11}\ \text{Pa} = 200\ \text{GPa}.$$

### Example

A 5.00 m long wire, of diameter 0.315 mm, is made of steel with a Young modulus of 200 GPa. Calculate its extension when a load of 10 N is applied to it.

### Answer

$$\sigma = \frac{F}{A} \text{ and } A = \pi r^2. \text{ So } \sigma = \frac{10\ \text{N}}{\pi \times \left(\dfrac{0.315 \times 10^{-3}\ \text{m}}{2}\right)^2} = 1.28 \times 10^8\ \text{Pa}$$

$$E = \frac{\sigma}{\varepsilon}, \text{ so } \varepsilon = \frac{\sigma}{E} = \frac{1.28 \times 10^6\ \text{Pa}}{200 \times 10^9\ \text{Pa}} = 6.4 \times 10^{-4},$$

$$E = \frac{\Delta l}{l_0}, \text{ so } \Delta l = \varepsilon l_0 = 6.4 \times 10^{-4} \times 5.00\ \text{m} = 3.2 \times 10^{-3}\ \text{m} = 3.2\ \text{mm}.$$

Remember that 1 cm$^2$ = 10$^{-4}$ m$^2$, so 1 N cm$^{-2}$ = 1 × 10$^4$ Pa = 10 kPa.

Notice that, in the answer to the example, the diameter of the wire is converted to m straightaway. This helps to avoid mistakes!

**Note**: It is quicker to use $E = \dfrac{Fl_0}{A\Delta l}$ if you can remember it. The tip is that, in calculating $E$ (a very big number), the very small numbers ($A$ and $\Delta l$) are on the bottom!

**Grade boost**

Remember to halve the diameter when using $A = \pi r^2$. Alternatively use

$$A = \pi \left(\frac{d}{2}\right)^2 \text{ or } A = \pi \frac{d^2}{4}$$

## 1.5.3 Work of deformation

We saw in Section 1.4.2 that, for a spring which obeys Hooke's law, the work, $W$, done in stretching the spring is given by

$$W = \tfrac{1}{2} Fx = \tfrac{1}{2} kx^2,$$

where     $k$ = the spring constant

And if the spring is elastic, i.e. the unloading curve is the same as the loading curve, the same formulae give us the elastic potential energy, also called the **strain energy**.

The same is also true for the same reason for a specimen of a *material* (e.g. a rod or wire) which obeys Hooke's law: the work done in stretching is equal to the area under the tension–extension graph; and the material can do the same work in relaxing.

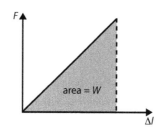

Fig. 1.5.4 Strain energy

It is often useful to consider the energy stored per unit volume, i.e. the strain energy density, of the material. This can be derived as follows:

The volume, $V$, of the material is given by $Al_0$ so the energy density is given by:

$$\frac{\tfrac{1}{2} F\Delta l}{Al_0} = \tfrac{1}{2} \frac{F}{A} \times \frac{\Delta l}{l_0} = \tfrac{1}{2} \sigma \varepsilon$$

The csa and length of a specimen are not strictly constant but the variation in $A$ and $l$ is so small that the unstrained values can be used. An interesting case is rubber: it does have large extensions but it also has a virtually constant volume when stretched.

Using the relationship $E = \dfrac{\sigma}{\varepsilon}$, the energy density per unit volume is given by

the three expressions: $\tfrac{1}{2} \sigma \varepsilon = \tfrac{1}{2} E\varepsilon^2 = \tfrac{1}{2} \dfrac{\sigma^2}{E}$.

**quickfire**

⑦ A load of 5 N extends a Hookean spring by 30 cm. What energy does the spring store?

**Grade boost**

The expressions for the work done (energy stored) per unit volume are not given in the Data Booklet:

$$\tfrac{1}{2} \sigma \varepsilon = \tfrac{1}{2} E\varepsilon^2 = \tfrac{1}{2} \frac{\sigma^2}{E}$$

**quickfire**

⑧ A 1.0 km long steel cable of cross-sectional area 20 cm$^2$ is extended by 1 m. Calculate the energy stored. Take the Young modulus of steel to be 200 GPa.

*Fig 1.5.5 Catapult*

>> **Pointer**
As dealt with in Section 1.5.5(c), rubber does not obey Hooke's law, so the 'Young modulus' is an average figure.

>> **Pointer**
In this example, we doubled the csa because of the two pieces of rubber but we could just as easily have calculated the energy in one band and multiplied this energy by 2.

**Key Terms**

**Crystalline** = consisting of a crystals; regular arrays of particles (usually ions).

**Polycrystalline** = consisting of a large number of interlocking crystals.

**Metal** = a condensed material (solid or liquid) in which the atoms have lost one or more electrons, to become positive ions, which are held together by the released 'delocalised' electrons.

## Example

The band of a catapult consists of two 10-cm long pieces rubber of cross section 4 mm × 4 mm. A 25 g stone is inserted, the band drawn back a distance of 10 cm and released. Taking the Young modulus of the rubber to be 20 MPa (see Pointer), calculate the speed that the stone will leave the catapult.

## Answer

csa of each piece of rubber = $(4 \times 10^{-3}$ m$)^2 = 1.6 \times 10^{-5}$ m$^2$.

So total csa = $3.2 \times 10^{-5}$ m$^2$.

∴ Volume of rubber = $0.1$ m $\times 3.2 \times 10^{-5}$ m = $3.2 \times 10^{-6}$ m$^3$

$\Delta l = 10$ cm; $l_0 = 10$ cm so $\varepsilon = 1$.

Energy stored per unit volume $= \frac{1}{2}E\varepsilon^2 = \frac{1}{2} \times 20 \times 10^6$ Pa $\times 1^2$

$= 1 \times 10^7$ J m$^{-3}$

∴ total energy stored = $1 \times 10^7$ J m$^{-3} \times 3.2 \times 10^{-6}$ m$^3$ = 32 J.

If all this energy is transferred to the stone, $E_k = \frac{1}{2} \times 0.025v^2 = 32$ J

$$\therefore v = \sqrt{\frac{2 \times 32}{0.025}} = 51 \text{ m s}^{-1} \text{ (2 s.f.)}$$

# 1.5.4 Different classes of solid materials

Materials may be classified as (poly)crystalline, amorphous or polymeric. These have very different tensile properties.

## (a) Crystalline materials

Crystals are solids with long-range order, i.e. the particles (usually atoms, molecules or ions) are in a regular arrangement, called a lattice. **Crystalline** engineering materials are most frequently **polycrystalline metals**. The lattice particles in metals are spherical metal ions, which in most metals are arranged hexagonally in the lattice planes.

In this arrangement each ion has the maximum possible number of near neighbours, so the potential energy of the hexagonal lattice is the lowest possible – Fig. 1.5.6 (a).

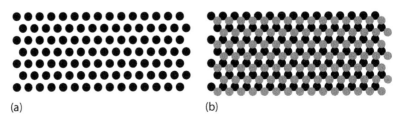

(a)                                          (b)

*Fig. 1.5.6 Metallic close-packed lattice*

The next lattice plane is also hexagonally arranged with each ion nestling into a gap between the ions in the place below – Fig. 1.5.6(b).

The same is true of the plane below, with the result that each ion is in contact with 12 near neighbours, which is the maximum possible. A piece of metal is generally not a single crystal but composed of a large number of crystals, each of which started to form separately from the molten state, resulting in random orientations of the crystal planes. This is illustrated schematically in Fig. 1.5.7. The diagram is not to scale – each of the crystals is about 50 μm across, which is about 6 orders of magnitude larger than the metal ions.

Fig 1.5.7 Polycrystalline metal: ions and grains

## (b) Amorphous materials

These materials have no long-range order. If a molecular material compound, e.g. silicon dioxide, is cooled rapidly from the molten state, its molecules don't have time to assume a crystal arrangement before they lose the ability to move. This results in a glass – as in Fig. 1.5.8.

Ceramics consist of molecules of metals and non-metals either covalently or ionically bonded. They are often partly crystalline within an amorphous matrix, unlike glasses which are entirely amorphous. Ceramics are not usually formed from a molten state.

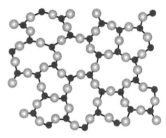

Fig.1.5.8 Glass structure

## (c) Polymeric materials

**Polymers** have very large molecules consisting of a very large number (~$10^5$ is not uncommon) of repeated sections. They are formed from **monomers**, which are molecules with one or more double bonds: the double bonds open giving rise to free bonds which can link the monomers together.

The simplest polymer is polythene (or polyethene): Figs 1.5.9 and 1.5.10.

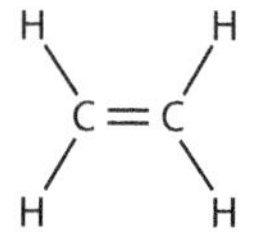

Fig. 1.5.9 Ethene – the monomer of polythene

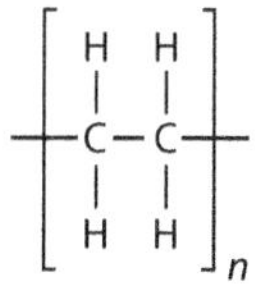

Fig. 1.5.10 Polythene molecule's repeat unit

Rubber is an almost ubiquitous polymer, polyisoprene: Figs 1.5.11 and 1.5.12

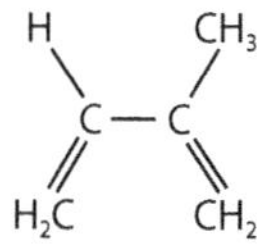

Fig. 1.5.11 Isoprene – the monomer of natural rubber

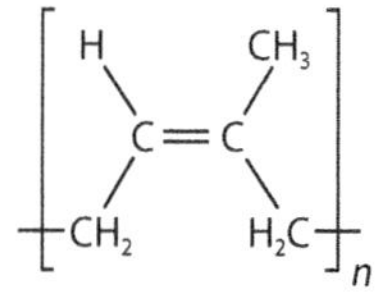

Fig. 1.5.12 Natural rubber, polyisoprene

The diagrams of polythene, rubber and their monomers are drawn as plane figures. In fact, they have a 3D structure: for a carbon atom with 4 single bonds, the angle between the bonds is about 110° ; single C–C bonds can rotate freely. These are important features explaining the mechanical properties of rubber.

**Key Terms**

**Polymer** = a material comprising large (macro-) molecules which consist of many repeat units.

**Monomer** = a molecule which can combine with other molecules to form a polymer.

**quicKpire**

(9) The formula of propene (propylene) is:

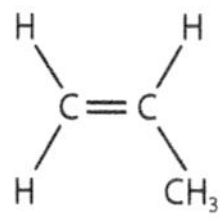

Draw the repeat unit for polypropene.

# 1.5.5 The stress–strain properties of ductile metals

## (a) The stress–strain relationships

Copper and steel are **ductile** materials. Features of their stress–strain curves are illustrated in Fig. 1.5.13

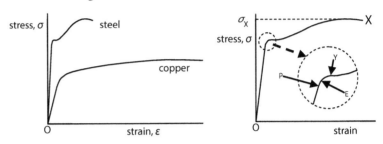

Fig. 1.5.13 $\sigma$ – $\varepsilon$ curves for copper and steel

The materials both have a linear elastic region for low stresses. The Young modulus is constant here.

At higher values of strain the graph curves to the right and material becomes **plastic**. The features on the graph for steel are more clearly defined:

P – limit of proportionality

E – the **elastic limit**

Y – the **yield point** and $\sigma_Y$ – the **yield stress**

$\sigma_X$ – the breaking stress

X – the breaking point

For low stresses, an increase in stress results in an increase in the separation of the lattice ions in the direction of the stress. When the stress is removed, the ions are pulled back by the metallic bonds. Thus the deformation is **elastic**. The explanation of **plastic** deformation is given in the next section.

## (b) Edge dislocations

Because the crystals grow randomly they are not perfect and very often an **edge dislocation** is produced [millions per crystal]. This is an extra part plane of ions. In Fig. 1.5.14 the dislocation is at X. If large enough forces are applied (so the stress exceeds the yield stress) as shown the dislocation will move irreversibly to the right i.e. the crystal will suffer permanent deformation. There are many animations of this – type *edge dislocation* into a search engine and choose a suitable video or sequence of images.

The movement of edge dislocations can cause large deformations in the following way: large stresses can cause a crystal plane to snap at a point of weakness (e.g. a missing ion) producing two edge dislocations, which migrate

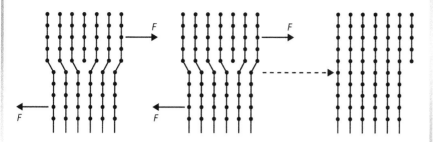

Fig. 1.5.14 Plastic deformation owing to the motion of an edge dislocation

in opposite directions. This makes the crystals elongate in the direction of the stress. Because there are large numbers of edge dislocations, quite large strains (20–30%) can be produced.

The edge dislocation cannot move beyond the grain boundary: the larger the grains, the greater the plastic strains. Foreign (impurity) atoms, grain boundaries and other dislocations impede the movement of edge dislocations and so stiffen and strengthen the material.

## (c) Ductile fracture

Fig. 1.5.15 represents **ductile fracture** with its typical necking and 'cup and cone' fracture. As the stress reaches $\sigma_X$, more and more edge dislocations are generated and migrate, causing the elongation. Because there is no increase in volume the cross-sectional area decreases, increasing the true stress at the neck, resulting in more dislocations in a runaway process. Flow marks are often seen in the necking region.

Fig. 1.5.15 Ductile fracture

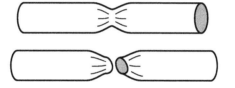

# 1.5.6 The stress–strain properties of brittle materials

## (a) The stress–strain relationship

Brittle materials are elastic; Hooke's law is obeyed for all stresses up to the breaking stress. This is illustrated in Fig. 1.5.16.

Note that 'elastic' does not mean highly extensible. The strain at the breaking point is of the order of 0.001 (i.e. 0.1%) for most brittle materials.

### Example

A rod of length 60 cm is made of glass with a Young modulus 70 GPa and a breaking stress of 100 MPa. Calculate the increase in length up to its breaking point.

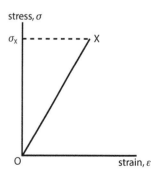

Fig. 1.5.16 $\sigma$–$\varepsilon$ graph for a brittle material

## Grade boost

Using your calculator:
Enter 100 M(Pa) as $100 \times 10^6$ and 70 G(Pa) as $70 \times 10^9$.

**Ultimate tensile strength (UTS)** = the maximum stress a material can withstand before breaking.

## ≫ Pointer

Brittle materials have no mobile edge dislocations because:

1. they are amorphous (no regular lattice), e.g. glasses, or
2. they are ionically or covalently bonded, e.g. ceramics, or
3. they are metallic but have very small crystals with large numbers of impurities, e.g. cast iron.

### quickfire

⑪ A glass fibre of diameter 0.12 mm just breaks when a mass of 650 g is hung from it. Its Young modulus is 80 GPa.

(a) Calculate the UTS.
(b) Calculate the strain at fracture.
(c) Sketch the stress–strain graph.

## ≫ Pointer

Masonry is very weak in tension – especially the mortar joints.
Old churches were built with towers at the corners of the walls. The weight of these helps keep the walls in compression.

---

**Answer**

$$E = \frac{\sigma}{\varepsilon}, \text{ so } \varepsilon_x = \frac{\sigma_x}{E} = \frac{100 \text{ MPa}}{70 \text{ GPa}} = 0.00143$$

$$\therefore \Delta l = \varepsilon_x l_0 = 0.00143 \times 60 \text{ cm} = 0.086 \text{ cm (2 s.f.)}$$

Note that we don't need to convert $l_0$ to m: the unit of $\Delta l$ is the same as the unit of $l_0$.

## (b) Brittle fracture

The tensile breaking stress of brittle materials is a lot lower than predicted from the strength of the bonds within the material, e.g. the theoretical UTS of glass ~6 GPa against the experimental value of up to 0.1 GPa for bulk glass. The material fractures because of the existence of microscopic cracks in the surface and these are responsible for the weakness of brittle materials under tension.

Fig. 1.5.17 shows a brittle sample with highly exaggerated crack. The pecked lines are so-called *stress lines*, which indicate how the tension is transmitted through the specimen. The stress lines are concentrated around the tip of the crack, magnifying the stress. There are no mobile edge dislocations to relieve the stress (see Pointer) so the crack breaks further at the tip, which produces an even higher stress at the new tip. The result is catastrophic failure: the crack propagates rapidly through the material.

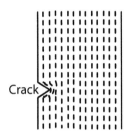

*Fig. 1.5.17 Brittle failure*

Experiments on thin glass fibres have shown that, if care is taken not to damage the surface, the UTS of a freshly drawn glass fibre increases with decreasing diameter. The thinner the glass fibre, the smaller the thermal stresses when it cools, so the problem with surface cracks is less. Very thin (~1 μm) glass fibres have strengths approaching the theoretical value.

## (c) The structural use of brittle materials

Brittle materials can be used in engineering if the propagation of cracks can be avoided. This is done in the following ways:

1. Concrete and brick structures are designed so that the brittle material is always *under compression*. In this way the cracks do not open up.
2. Pre-stressed concrete is made by pouring the concrete around steel rods under tension. The rods are slackened off when the concrete has cured, putting the concrete under compression.
3. Pre-stressed glass is manufactured so that the surface is under compression. This is done by a rapid cooling of the surface – the surface then sets first and the later cooling of the core puts the surface under compression. A tension can be applied without the crack-bearing surface being put under tension.

# 1.5.7 The stress–strain properties of rubber

The stress–strain and load–extension graphs have the following features:

- Non-linear: steep → less steep → very steep; Hooke's law is approximately obeyed only for very low stresses.
- Large strains: up to ~5 (or 500%) depending on the type of rubber.
- The stress needed to stretch is low, i.e. the Young modulus is very low.
- Loading and unloading curves different: called **elastic hysteresis**.

Because the area under a load–extension curve is the work done, the work done *by* the rubber band in contracting is less than the work done *on* the rubber band in stretching.

The area *between* the graphs represents the energy dissipated in moving once around the hysteresis loop. It manifests itself as random vibrational energy of the rubber molecules.

Why does rubber behave like this?

1. The C–C bonds in the long chains can rotate;
2. Successive C–C bonds are at an angle (~110°) to each other.

A rubber molecule in the unstressed state is naturally tangled up: it is very unlikely to form in a linear state [see Fig. 1.5.19 – remember it is 3D]. Applying a small longitudinal force rotates the bonds and straightens out the molecules: no bonds are stretched; large extensions are produced. The force works against the thermal motions of the molecules which tend to pull the ends in. When the force is relaxed, the natural vibration of the molecules tangles up the long chains again. Because of the work done, the molecules end up vibrating more, i.e. energy is dissipated.

The energy losses in hysteresis can be *useful*, e.g. in shock absorbers. It can be a *nuisance*, e.g. the rolling resistance in car tyres. It can be reduced by introducing cross-linkages between molecules (or different parts of the same molecule) in the process called *vulcanisation*.

**Key Term**

**Elastic hysteresis** = when a material such as rubber is put under stress and then relaxed, the stress–strain graphs for increasing and decreasing stress do not coincide but form a loop.

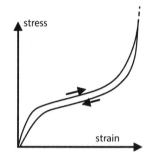

Fig. 1.5.18 Stress-strain curve for rubber

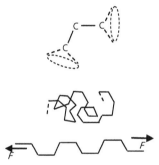

Fig. 1.5.19 Rotating bonds in rubber molecules

# 1.5.8 Specified practical work

## (a) Determination of the Young modulus for the material of a wire

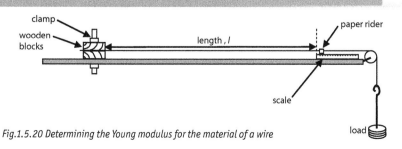

Fig.1.5.20 Determining the Young modulus for the material of a wire

## >> Pointer

The analysis of the data is as follows:

$E = \dfrac{Fl_0}{A\Delta l}$, so $F = \dfrac{EA}{l_0}\Delta l$,

where $A = \pi \dfrac{d^2}{4}$.

So the gradient, $m$, of the graph is given by

$m = \dfrac{EA}{l_0}$. $\therefore E = \dfrac{l_0 m}{A}$

A long piece of wire is used, with $l$ typically about 2 m. The blocks are to prevent damage to the wire.

The original length $l_0$ is measured using a metre rule (giving an uncertainty of ~ 0.1 %). The extension is measured using the paper rider (or a paint blob) and the mm scale – this is the least precise part of the experiment. For accurate work a travelling microscope can be used. The tension is determined from the mass of the load – typically increased in 0.1 kg steps – and use of $W = mg$. After the maximum load is reached, the load is decreased and the mean of the $\Delta l$ values for each of the two values of load is calculated.

The diameter of the wire is determined using a micrometer / digital calliper (giving an uncertainty of 0.01 mm, i.e 3% for a 0.3 mm diameter wire).

A graph is plotted of $F$ against $\Delta l$. The value of the Young modulus is calculated from the gradient as shown in the Pointer.

*Sample data:* Copper wire; $l_0 = 2.105 \pm 0.001$ m; $d = 0.38 \pm 0.01$ mm; gradient = $6200 \pm 300$ N m$^{-1}$.

*Calculation:* Gradient $m = \dfrac{EA}{l_0}$, so $E = \dfrac{l_0 m}{A} = \dfrac{2.105 \text{ m} \times 6200 \text{ N m}^{-1}}{\pi \left(\dfrac{0.38 \times 10^{-3}}{2}\right)^2} = 115$ GPa

*Uncertainty:* % uncertainties: $d$ – 2.6%; gradient – 4.8%; $l_0$ – 0.05% [ignore].

So total uncertainty = 2.6% + 4.8% = 7.4%. 7.4% of 115 = 8.5

So $E = 115 \pm 9$ GPa

## (b) Investigation of the force–extension relationship for rubber

This experiment is carried out as for the spring in Section 1.5.1. The extensions are measured for increasing loads until the rubber band shows very little additional extension. The load is then gradually decreased and the extensions measured. A load–extension graph of the same shape as the graph in Fig. 1.5.2 is obtained, clearly showing elastic hysteresis.

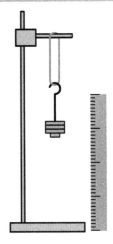

*Fig. 1.5.21 Force-extension for rubber*

# Extra questions

1. Express the unit of the spring constant, $k$, in terms of the base SI units.

2. The unloaded length of a spring is 15.3 cm. When a 500 g mass is suspended from it, its length is 23.7 cm. Calculate the spring constant.

3. Two springs of constant 200 N m$^{-1}$ and 300 N m$^{-1}$ at joined end to end. Calculate the strain energy when the springs are put under a tension of 150 N.

4. A rod, X, has a cross-sectional area $A$. Rod Y has the same length and is made of the same material. Its csa is $2A$.
   (a) Compare the elastic potential energies if the rods are put under the same tension.
   (b) Compare the elastic potential energies if $X$ is put under a tension $2T$ and Y is put under a tension $T$.

5. An engineer's data book gives the following data for copper:
   Young modulus = 117 GPa; Yield stress = 75 MPa;
   UTS = 150 MPa; breaking strain = 0.45.
   A copper wire has diameter 3.25 mm and length 1 km.
   (a) The wire is subject to a tension which is 90% of the yield stress. Calculate (i) the tension, (ii) the extension and (iii) the strain energy.
   (b) Sketch the load–extension curve and use it to estimate the work done in stretching it to its breaking point.

6. A student obtained the following results for a piece of copper wire:
   Diameter of wire, $r$ = 0.272 mm; $l_0$ = 3.43 m

   | Load / g | 50 | 100 | 150 | 200 | 250 | 300 | 350 | 400 |
   |---|---|---|---|---|---|---|---|---|
   | Extension / mm | 0.20 | 0.51 | 0.78 | 1.03 | 1.29 | 1.50 | 1.68 | 1.98 |

   Use these results to determine the Young modulus of the copper.

>> *Pointer*
From the wavelengths of absorption lines in the Sun's spectrum we know there are elements (such as sodium) with Z > 6 in the Sun's atmosphere. These elements are formed only in extreme events such as supernovae, so we deduce that the Sun doesn't date from the very early universe!

# 1.6 Using radiation to investigate stars

We can learn much about a star from its **spectrum** – how its e-m radiation energy is distributed across infrared, visible and ultraviolet wavelengths. We find this out using a telescope, a diffraction grating and suitable detectors. Multiwavelength astronomy (Section 1.6.5) vastly extends the range of wavelengths detected and analysed.

## 1.6.1 A star's spectrum

The star's spectrum consists of:

- A continuous spectrum of radiation arising from the dense, opaque gas of the star's surface.

- A superimposed line absorption spectrum (Section 2.7.8) due to atoms in the star's atmosphere, through which the radiation must pass.

    From the wavelengths of the dark lines physicists can identify the absorbing atoms responsible for them. See Pointer for example.

## 1.6.2 The continuous spectrum

This approximates to what physicists call a **black body spectrum**.

A black body is an ideal surface that absorbs all the e-m radiation falling on it. At any given wavelength we find that the best absorbers are also the best emitters. In particular, no surface emits more radiation per $m^2$, purely because it is hot, than a black body at the same temperature.

For a black body the *spectral intensity* or *spectral radiance*[1] (the power emitted per unit area per unit interval of wavelength) depends on the wavelength as shown in the spectra (drawn for three temperatures).

>> *Pointer*
Because it is emitting so strongly, a star doesn't *look* black. Nonetheless it is absorbing (almost) all radiation from external bodies that falls on it – we just don't notice this happening!

>> *Pointer*
Make sure that you can sketch the shape of a black body spectrum. Note the zero gradient at the origin.

### quickpire

① In which regions of the e-m spectrum are the wavelengths of peak emission for black bodies at temperatures of:
(a) 5000 K, (b) 4000 K, (c) 3000 K ?

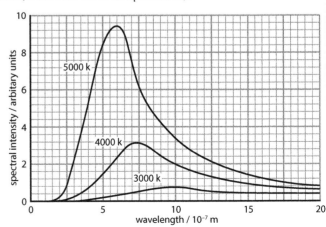

*Fig 1.6.1 Three black body spectra*

[1] You are not required to know the terms spectral intensity or spectral radiance.

### Example

What colour is a star whose surface temperature is 3000 K?

### Answer

The wavelength of peak emission is about $10 \times 10^{-7}$ m (1000 nm), which is in the infrared region. But we're interested only in wavelengths in the visible! The spectral intensity falls markedly as we go from 700 nm (deep red) to 400 nm (violet), so the star will be reddish (Section 2.7.2).

## 1.6.3 Wien's displacement law for black body radiation

The wavelength, $\lambda_P$, at which the black body spectrum (spectral intensity against wavelength) has its peak is inversely proportional to the black body's kelvin temperature, $T$.

That is $\lambda_P = \dfrac{W}{T}$ in which $W$ is *the Wien constant*. $W = 2.90 \times 10^{-3}$ K m.

In Fig 1.6.1 the wavelength, $l_P$, of peak spectral intensity is *displaced* further and further to the left as the temperature increases. You should evaluate $T\lambda_P$ for all three spectra, to check agreement with Wien's law.

## 1.6.4 Stefan's law for black body radiation

The total power, $P$, of electromagnetic radiation emitted from area, $A$, of a black body at kelvin temperature $T$ is given by

$$P = \sigma A T^4$$

in which $\sigma$ is called *the Stefan constant* $\sigma = 5.67 \times 10^{-8}$ W m$^{-2}$ K$^{-4}$.

### Example: applying the black body laws to the star Arcturus

Arcturus is a bright orange star for which $\lambda_P$ is measured to be 674 nm. The star is known to be $3.47 \times 10^{17}$ m from Earth, and the intensity of radiation received on Earth from it is $3.09 \times 10^{-8}$ Wm$^{-2}$. Calculate:

(a)  The star's temperature

(b)  Its luminosity

(c)  Its radius.

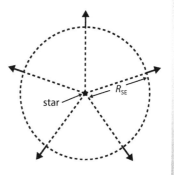

*Fig 1.6.2 Intensity*

### Answer

(a)  We can find the star's temperature using Wien's law...

$$T = \frac{W}{\lambda_p} = \frac{2.90 \times 10^{-3} \text{ K m}}{674 \times 10^{-9} \text{ m}} = 4300 \text{ K.}$$

**Key Terms**

**Wien's displacement law**

$\lambda_p = \dfrac{W}{T}$  [$W = 2.90 \times 10^{-3}$ K m]

Law stated in main text.

**Stefan's law**

$P = \sigma A T^4$  [$\sigma = 5.67 \times 10^{-8}$ W m$^{-2}$ K$^{-4}$]

Law stated in main text. Sometimes called *the Stefan-Boltzmann law*.

[$T$/K = $\theta$/°C + 273.15 relates **kelvin temperature**, T, to Celsius temp., $\theta$.]

A star's **luminosity** is the total power of e-m radiation it emits. Unit: W

The **intensity** of e-m radiation is the energy passing per unit time, per unit area, through an area normal to the propagation direction. Unit: W m$^{-2}$

**Inverse square law** for radiation intensity at distance $R_{SE}$ from star...

$$\text{Intensity} = \frac{\text{Luminosity}}{4\pi R_{SE}^2}$$

Doubling distance quarters intensity, and so on. The law assumes no **absorption** of radiation as it travels.

(b) A star's **luminosity** is the total power, $P$, of e-m radiation it emits. At the distance $R_{SE}$ of the Earth from the star, this power is passing through a spherical surface of area $4\pi R_{SE}^2$.

So the **intensity** of e-m radiation (the power per unit area of surface crossed) is

$$\text{Intensity, } I = \frac{P}{4\pi R_{SE}^2} \quad \text{so} \quad P = I \times 4\pi R_{SE}^2$$

Here, $P = 3.09 \times 10^8$ m$^{-2}$ $\times 4\pi \times (3.47 \times 10^{17}$ m$)^2 = 4.68 \times 10^{28}$ W

(c) Knowing the star's luminosity and its temperature, we can now find its surface area, $A$, using Stefan's law. Because the star will be almost spherical, we can put $A = 4\pi r^2$ and so find the radius $r$ directly...

$$\text{So } r = \sqrt{\frac{P}{4\pi\sigma T^4}} = \sqrt{\frac{4.68 \times 10^{28} \text{ W}}{4\pi \times 5.67 \times 10^{-8} \text{ W m}^{-2} \text{ K}^{-4} (4300 \text{ K})^4}}$$

$$= 1.39 \times 10^{10} \text{ m}$$

# 1.6.5 Multiwavelength astronomy

Multiwavelength astronomy is the detection and analysis of e-m radiation from stars and other objects in space, over wavelengths ranging from radio waves to gamma rays (Section 2.7.2).

## (a) Astronomy from outside the Earth's atmosphere

Our atmosphere absorbs gamma rays, X-rays, most ultraviolet (thanks to the ozone layer) and all but certain narrow wavelength bands of infrared.

What has made true multiwavelength astronomy possible is our relatively new ability to put equipment for gathering and analysing e-m radiation into space stations or 'observatories' orbiting the Earth – beyond its atmosphere. There are now many such observatories, each specialising in a particular region of the spectrum (for example gamma or X-ray).

## (b) Thermal sources of radiation

Usually bodies in space emit radiation because they are hot, and Wien's law (Section 1.6.3) gives a rough guide to the temperature of a body emitting radiation in a particular spectral region.

### Example

What would be a typical temperature for a body that emits mainly in the X-ray region?

### Answer

Referring to Fig 2.7.1, a typical X-ray wavelength is $1 \times 10^{-11}$ m, and if this were $\lambda_P$ for a black body its temperature would be

$$T = \frac{W}{\lambda_p} = \frac{2.90 \times 10^{-3} \text{ K m}}{1 \times 10^{-11} \text{ m}} \approx 3 \times 10^8 \text{ K}.$$

Hot X-ray sources include material attracted by galaxy clusters and colliding at very high kinetic energy with material already captured. See also Extra question 4.

## (c) Looking into the past and at different ages

Multiwavelength astronomy can be at its most revealing when the same area of space is observed at different wavelengths. For example, the Andromeda galaxy (aka M31) has been studied in the infrared, showing regions of star formation and the overall galactic structure of rings and spiral arms. Stars in their main phase show up well in visible studies. X-ray studies of M31 reveal intense sources, mainly associated with the capture of material by high mass stars that have used up their original nuclear fuel and evolved into black holes or neutron stars. Supernovae remnants in M31 are also powerful X-ray emitters. Hence, different ages of stars can be investigated depending on which wavelengths you choose to analyse – from radio waves for forming stars, through visible for main sequence to X-ray for dead giant remnants.

When we detect e-m radiation from distant objects in space we are 'looking into the past'. For example, M31 is 2.5 million light years ($2.3 \times 10^{22}$ m) away, so we are observing what it was like 2.5 million years ago!

### quickfire

⑦ R136a1 (see Quickfire 6) is $1.54 \times 10^{21}$ m away. The intensity of radiation from it is $1.0 \times 10^{-10}$ W m$^{-2}$. Calculate:
(a) its luminosity
(b) its diameter.

### ≫ Pointer

The most distant galaxy observed so far was formed around 700 million years after the Big Bang. This galaxy is around 13.1 billion years old and the images of it are formed from photons that have been travelling undeflected for 13.1 billion years.

## Extra questions

1. The cosmic microwave background has the spectrum of a black body at 2.7 K, with $\lambda_p$ at 1.06 mm. This radiation is the cooled remnant of black body radiation from an early epoch when the temperature was around 3000 K. In which region or regions of the e-m spectrum did this radiation lie?

2. One of the hottest stars known is HD93129A in the Carina nebula. Its continuous spectrum is shown on page 66.
   (a) (i) Name the region of the electromagnetic spectrum in which the peak emission lies, and explain why the star's colour is blue.
       (ii) Show that the star's temperature is about $5 \times 10^4$ K.
   (b) (i) Use the following data to calculate the luminosity of HD93129A:
           Distance: $7.10 \times 10^{-19}$ m
           Intensity of radiation at Earth: $3.33 \times 10^{-8}$ W m$^{-2}$.
       (ii) Show that its luminosity is approximately $5 \times 10^6 \, P_{sun}$ in which $P_{sun}$ is the Sun's luminosity ($3.84 \times 10^{26}$ W).
       (iii) Calculate HD93129A's diameter.

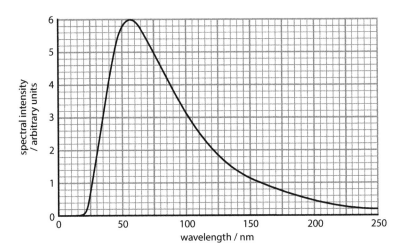

3. Using the definition of *spectral intensity* in Section 1.6.2:

(a) give the SI unit of spectral intensity.

(b) state what quantity the area under the graphs in Fig. 1.6.1 represents.

4. *Neutron stars* are very small, dense, 'dead' stars. Sometimes they acquire a very hot 'active' outer layer which radiates as a black body. One such star has a radius of 11 km, and radiates at a temperature of $2.5 \times 10^7$ K.

(a) Calculate the wavelength of greatest spectral intensity of its emitted radiation, and name the region of the e-m spectrum in which it lies.

(b) Calculate the star's luminosity.

(c) The outer layer of the star expands rapidly and cools. The luminosity remains roughly constant. Estimate the temperature of the outer layer when its surface area has doubled.

5. (a) Explain why a supermassive black hole known as a quasar might be best imaged using a gamma ray telescope rather than a visible light telescope. The light from many quasars is observed after it has travelled up to 10 billion light years. How and why might this affect your answer?

(b) A more massive main sequence star will be hotter than a smaller star because the increased density at its core due to gravity leads to a far higher rate of fusion reactions. Use this to explain why the colour of a main sequence star can give you information about its mass.

(c) State which region of the e-m spectrum is best for analysing the following objects:

(i) Planets,

(ii) Regions where cosmic rays collide with hydrogen nuclei,

(iii) Very cold areas of the interstellar medium,

(iv) A region of dense nebula that is soon to form a star,

(v) The cosmic microwave background radiation,

(vi) Stellar coronas (high energy discharge from stars).

# 1.7 Particles and nuclear structure

This topic is the most fundamental of all subjects and is the underlying foundation of the whole universe. Essentially, this section introduces the building blocks of matter (quarks and leptons) and their place in the standard model. The relevance and importance of the four fundamental force laws are also explained with reference to particle interactions.

## 1.7.1 What is matter?

The fundamental particles in the universe, according to particle physicists' *Standard Model* are summarised in Fig. 1.7.1

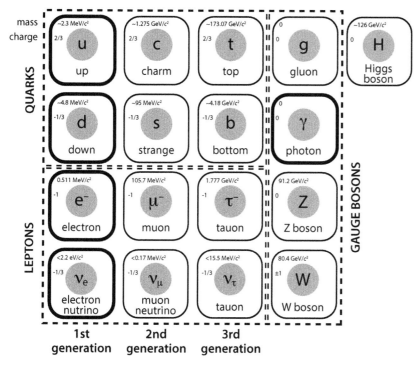

Fig. 1.7.1 The standard particle model

This equivalent of the chemists' periodic table was not all discovered in one go. A brief timeline is given in the Pointer. All normal matter in the universe is composed of the particles in the first three columns (the leptons and quarks), their interactions controlled by the *gauge bosons* (gluons, photons, W and Z bosons), with the newly discovered Higgs boson being responsible for their masses.

**For the exam**, you need to know about the particles outlined heavily – the first generation leptons and quarks and the photon. You should also know that there are three generations – the second and third generations were discovered as more and more powerful particle accelerators became available.

**Key Terms**

**Antiparticle** = a subatomic particle with the same mass as a given particle but the opposite charge.

**Quark** = a fundamental particle that experiences the strong nuclear force.

**Lepton** = a fundamental particle that experiences the weak nuclear force (but not the strong nuclear force).

**Grade boost**

You must ensure that you know all these definitions and which particles feel which forces.

**Pointer**

Electron 1897 (JJ Thompson)
Proton accepted 1920s
Neutron 1932 (James Chadwick)
Electron neutrino predicted 1930s, detected 1956
Quarks predicted 1960s detected 1969
W and Z bosons 1983 (CERN)
Higgs boson 2013 (CERN)

## Grade boost

Make sure you know all the particles, antiparticles and their symbols in Table 1.7.1.

**QUICKFIRE**

① The symbol for the strange quark is s. Write down the symbol for its antiparticle and suggest a name for it.

>> **Pointer**

For most particles, the symbol for the antiparticle is the same as for the 'normal' particle with a bar above it. The exceptions are $e^+$, $\mu^+$ and $\tau^+$ which are the antiparticles for $e^-$, $\mu^-$ and $\tau^-$ respectively.

>> **Pointer**

In different areas of physics, people use different symbols for the electron:

$e$, $e^-$, $_{-1}^{0}e$ and $\beta^-$

Similarly, for the positron:

$e^+$, $_{1}^{0}e$ and $\beta^+$

are used in different contexts.

>> **Pointer**

The simplest type of annihilation is when a particle and antiparticle produce two photons but annihilation interactions can be more complicated, e.g.
$p + \bar{p} \rightarrow 2\pi^+ + 2\pi^- + \pi^0$

>> **Pointer**

A-level students should be able to use $E = mc^2$ to show that the energy of the photons produced from electron-positron annihilation is as given in Fig 1.7.2.

# 1.7.2 Antimatter

All quarks and leptons have their equivalent antiparticles. These antiparticles have the same properties as the corresponding particle except that the charge is opposite. For example, an electron has an antiparticle called a positron (or anti-electron, $e^+$ or $\beta^+$). The mass of a positron is the same as that of an electron ($9.11 \times 10^{-31}$ kg) but its charge is opposite ($+1.60 \times 10^{-19}$ C). The antiparticle of the proton, the antiproton, ($\bar{p}$) has the same mass as a proton but its charge is opposite. Likewise, an anti-up quark ($\bar{u}$) has the same properties as an up quark (u) but its charge is the opposite

$(-\frac{2}{3}e$ instead of $+\frac{2}{3}e)$.

Note that even the electron neutrino ($\nu_e$) has a corresponding antiparticle even though it is neutral. The antiparticle of the electron neutrino ($\nu_e$) is the anti-electron neutrino ($\bar{\nu}_e$).

Table 1.7.1 contains particles and antiparticles whose names and symbols you should know.

| Particle | Symbol | Antiparticle | Symbol |
|----------|--------|--------------|--------|
| Up quark | u | Anti-up quark | $\bar{u}$ |
| Down quark | d | Anti-down quark | $\bar{d}$ |
| Proton | p | Antiproton | $\bar{p}$ |
| Neutron | n | Antineutron | $\bar{n}$ |
| Electron | $e^-$ (or just e ) | Positron or anti-electron | $e^+$ |
| Electron neutrino | $\nu_e$ | Anti-electron neutrino | $\bar{\nu}_e$ |

*Table 1.7.1 Particles and symbols you need to know*

# 1.7.3 Annihilation or what happens when antimatter and matter meet?

This is the process whereby all or nearly all mass is lost and converted to energy according to Einstein's famous equation $E = mc^2$. An example of annihilation that occurs naturally in the atmosphere due to cosmic rays is electron–positron annihilation.

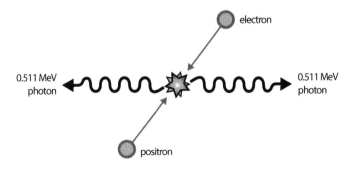

*Fig. 1.7.2 $e^+e^-$ annihilation*

Here, when the positron encounters an electron the result is the complete loss of mass of the original particles. This mass loss is converted into energy in the form of the two photons that are emitted in opposite directions (due to conservation of momentum).

# 1.7.4 Quarks, hadrons, baryons and mesons

Although a form of matter called a 'quark soup' probably existed in the extremely high temperatures shortly after the Big Bang, nowadays quarks are never seen in isolation. They always appear in **hadrons**, which are composite particles and can be either of two types:

**Baryons** = three quarks (qqq) **Mesons** = quark-antiquark pairs ($q\bar{q}$)

## (a) Baryons ('heavy ones') and antibaryons

By definition, a baryon is a composite particle made up of any 3 quarks and an antibaryon is a composite particle made up of any 3 antiquarks. There is only one stable baryon – the proton. A free neutron has a half-life of around 10 minutes but neutrons in the nuclei of atoms are usually stable. Protons are so stable that the free-proton decay has never been detected. All this is excellent news for matter because it is mostly made of protons and neutrons but what is the quark make-up of protons and neutrons?

Table 1.7.2 gives you all the properties of the up and down quarks that you require. In order to build a **proton** out of the u and d quarks we must arrive at a final charge, $Q$, of $+e$ (or +1) and a baryon number, $B$, of 1. (It's one baryon – so its baryon number is 1.)

| Particle (symbol) | quarks | |
| --- | --- | --- |
| | Up (u) | Down (d) |
| Charge, $Q/e$ | $+\frac{2}{3}$ | $-\frac{1}{3}$ |
| Baryon number, $B$ | $+\frac{1}{3}$ | $+\frac{1}{3}$ |

Table 1.7.2 Quark properties

The only possible combination of three u and d quarks to produce these properties is uud.

because $Q = \frac{2}{3} + \frac{2}{3} - \frac{1}{3} = 1$ and $B = \frac{1}{3} + \frac{1}{3} + \frac{1}{3} = 1$

The concept of baryon number is very important as it is a conserved quantity, i.e. the total baryon number never changes in particle interactions (see Section 1.7.5)

### ≫ Pointer

The $\Delta^+$ has the same quark composition as a proton. So how is it a different particle? All the $\Delta$s are 30% heavier than protons and neutrons. Most of the mass of particles comes from factors other than the individual quark masses.

### ≫ Pointer

All quarks have a baryon number of $\frac{1}{3}$; antiquarks have a baryon number of $-\frac{1}{3}$

### quickfire

③ A first generation antiparticle has quark composition $\overline{uud}$. State its charge and identify two possible names for it.

### quickfire

④ State the quark composition of $\overline{\Delta^-}$.

### ≫ Pointer

As well as an overall baryon number it is useful to define individual quark numbers: the up number ($U$) and down number ($D$). We'll return to this in Section 1.7.6.

|  | u | d |  |
|---|---|---|---|
| $\overline{u}$ | $u\overline{u}$ $Q=0$ $\pi^0$ | $u\overline{d}$ $Q=-1$ $\pi^-$ | Composition charge symbol |
| $\overline{d}$ | $u\overline{d}$ $Q=1$ $\pi^+$ | $d\overline{d}$ $Q=0$ $\pi^0$ |  |

Table 1.7.5 Pion composition

## 1st generation baryons – the complete set:

| particle | p | n | $\Delta^{++}$ | $\Delta^+$ | $\Delta^0$ | $\Delta^-$ |
|---|---|---|---|---|---|---|
| structure | uud | udd | uuu | uud | udd | ddd |
| lifetime | stable | ~10 min | \multicolumn{4}{c}{$5.6 \times 10^{-24}$ s} |

Table 1.7.3 First generation baryons

Note that, oddly, p and $\Delta^+$ have the same quark structure, as do n and $\Delta^0$ (see Pointer). The $\Delta$s are the delta baryons: delta double plus, delta plus, delta zero and delta minus.

## Antibaryons

To produce antibaryons as well we need the slightly more complete quark-property table, Table. 1.7.4

| Particle (symbol) | \multicolumn{2}{c}{quarks} | \multicolumn{2}{c}{antiquarks} | | |
|---|---|---|---|---|
|  | Up (u) | Down (d) | Anti-up ($\overline{u}$) | Anti-down ($\overline{d}$) |
| Charge, $Q/e$ | $\frac{2}{3}$ | $-\frac{1}{3}$ | $-\frac{2}{3}$ | $\frac{1}{3}$ |
| Baryon number, $B$ | $\frac{1}{3}$ | $\frac{1}{3}$ | $-\frac{1}{3}$ | $-\frac{1}{3}$ |

Table 1.7.4 $q\overline{q}$ properties

Notice that the $Q$ and $B$ values of the antiquarks are exactly what you would expect them to be, i.e. equal but opposite.

### Example

Find the (anti)quark composition of the antineutron.

### Answer

For the antineutron, $Q = 0$ and $B = -1$.

It must be made of 3 antiquarks (like all antibaryons): $-\frac{1}{3} - \frac{1}{3} - \frac{1}{3} = -1$.

So $\overline{n} = \overline{udd}$ because $Q = -\frac{2}{3} + \frac{1}{3} + \frac{1}{3} = 0$.

## (b) Mesons ('middle ones')

By definition, a meson is a composite particle made up of one quark and one antiquark. You only have to answer questions on mesons with combinations of u, d, $\overline{u}$ and $\overline{d}$ quarks. There are just 4 possibilities for these mesons, which are known as pions. Their make-up is shown in Table 1.7.5, the pion 'Punnett square'.

The pions are known as $\pi$-plus ($\pi^+$), pi-minus ($\pi^-$) and pi-zero ($\pi^0$), with the superscript indicating their charge.

You should note two things about pions:

1. Because they are composed of a quark and antiquark pair, their baryon number, $B = 1 - 1 = 0$. **This is true of all mesons, not just pions** and will turn out to be important when we look at particle interactions.

2. The neutral pion, $\pi^0$, seems to have two possible make-ups: $u\bar{u}$ and $d\bar{d}$. The truth is even stranger (see Pointer) but this is all you need to know.

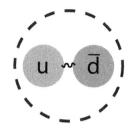

*Fig. 1.7.3 Structure of a $\pi^+$*

Just like the baryons, there are also heavier versions of the pions, called the rho mesons ($\rho^+$, $\rho^0$ and $\rho^-$) and the omega meson ($\omega^0$), so don't be surprised if the examiner throws them at you in the exam!

> **Pointer**
> The $\pi^0$ is actually a combination of $u\bar{u}$ and $d\bar{d}$. It could be found in either state (like Schrödinger's cat!).

**quicKpire**

⑤ Explain why the antiparticle of the $\pi^+$ meson is the $\pi^-$ meson.

> **Pointer**
> The $\rho$ and $\omega$ mesons have the same quark structure as the equivalent pions.

## 1.7.5 Leptons ('light ones')

Unlike hadrons (baryons and mesons) but like quarks, electrons and neutrinos are **elementary particles**. However, unlike quarks, they do not combine to produce composite subatomic particles[1] – they always exist separately. The important properties of leptons are:

1. Charge, $Q$: electrons (and muons and tauons) have $Q = 1$; all the neutrinos are uncharged.

2. Lepton number: electrons ($e^-$) and electron neutrinos ($v_e$) have a lepton number, $L = 1$; for positrons ($e^+$) and anti-electron neutrinos ($\bar{v}_e$), $L = -1$.

What about the baryon number? Leptons (like mesons) are not baryons, their baryon number is therefore 0. Similarly the lepton number of all baryons and mesons is 0.

| Symbol | Q/e | L |
|--------|-----|---|
| $e^-$ | −1 | 1 |
| $v_e$ | 0 | 1 |
| $e^+$ | 1 | −1 |
| $\bar{v}_e$ | 0 | −1 |

*Table 1.7.6 Lepton properties*

> **Pointer**
> In fact, the different generations of leptons have individual lepton numbers, $L_e$, $L_\mu$ and $L$. You only need to know about $L_e$, so we'll just call it $L$ here.

## 1.7.6 Conservation laws

We're almost ready to consider particle interactions. Before doing this, there are three fundamental conservation laws that you need to understand and apply, two of which are new (see also Grade boost).

(i) Conservation of charge, $Q$;

(ii) Conservation of baryon number, $B$;

(iii) Conservation of (electron) lepton number, $L$.

These three laws apply to **all** particle interactions and you will use them to analyse various interactions and identify 'unknown' particles.

We'll look at how all these conservation laws apply to the $\beta^+$ decay of fluorine-18. Looking at the whole nucleus the reaction is:

$$^{18}_{9}F \rightarrow\ ^{18}_{8}O + {}^{0}_{1}\beta + X,$$

where X is a particle whose identity is, for the moment, a mystery!

⚹ **Grade boost**
In fact, there is a fourth conservation law: the conservation of mass-energy. This is just the conservation of energy taking into account the energy locked up in the mass of the particles according to $E = mc^2$.

> **Pointer**
> Note that the symbols $^{18}_{8}O$ and $^{18}_{9}F$ represent *nuclei* not *atoms* in this context.

---

[1] Electrons do link with protons and neutrons in atoms of course.

## Grade boost

(For A level students) A neutron is more massive than a proton, so how can a proton decay into a neutron? Answer: An isolated proton cannot – it is stable. But the $^{18}_{8}\text{O}$ is more tightly bound than the $^{18}_{9}\text{F}$. The energy this releases allows it to produce a neutron and a beta particle as well as particle X.

This reaction happens when one proton in the nucleus converts to a neutron with the emission of a positron and X. So we'll rewrite this in particle physics notation and apply the conservation laws:

$$p \rightarrow n + e^+ + X$$

Charge:    $Q$:    $1 = 0 + 1 + Q_X$    $\therefore Q_X = 0$

(So we know that X is uncharged.)

Baryon number  $B$:    $1 = 1 + 0 + B_X$    $\therefore B_X = 0$

(So now we know that X is a not baryon and that it is uncharged. So it could be an uncharged lepton or meson. But we're not finished yet.)

Lepton number   $L$:    $0 = 0 + (-1) + L_X$   $\therefore L_X = 1$

So X is an uncharged lepton with a lepton number of 1, i.e. it is an electron neutrino, $\nu_e$.

We'll come back to β decay (β⁻ as well as β⁺) when we've looked at the four forces.

## » Pointer

We can ignore the gravitational interaction as far as particle physics is concerned, unless we are just outside the event horizon of a black hole! It is so much weaker than the other forces.

# 1.7.7 The four forces of the universe

Physicists recognise four fundamental interactions between material particles. The syllabus kindly provides you with this rather useful table, which you should learn:

| Interaction | Experienced by | Range | Comments |
|---|---|---|---|
| Gravitational | All matter | Infinite | Very weak – negligible except between large objects such as planets |
| Weak | All leptons, all quarks, so all hadrons | Very short | Only significant when the e-m and strong interactions do not operate |
| Electro-magnetic (e-m) | All charged particles | Infinite | Also experienced by neutral hadrons, as these are composed of charged quarks |
| Strong | All quarks, so all hadrons | Short | *Binds quarks together in hadrons. Binds protons and neutrons in the nucleus.* |

Table 1.7.6 The four fundamental forces

The text in italics does not appear in the syllabus but has been added for completeness. These forces govern all particle interactions including:

1. Elastic collisions, e.g. $e^- + e^- \rightarrow e^- + e^-$
2. Particle annihilation and production, e.g. $p + \bar{p} \rightarrow 2\pi^- + 2\pi^+ + \pi^0$
3. Particle decay, e.g. $n \rightarrow p + e^- + \bar{\nu}_e$

Here are some tell-tale signs about the interaction which governs a reaction:

## (a) Strong interaction

- Particle decays have very short lifetimes, typically ~$10^{-24}$ s.
- The interaction is very likely to happen when particles collide, e.g. interaction 2.
- All particles involved are hadrons.
- There is no change in quark flavour.

This last point needs examining. We keep a check on the total u-quark number, $U$, and the total d-quark number, $D$. As with the lepton and baryon numbers, the anti-u has $U = -1$. Look at equation 2 above:

$$p + \bar{p} \rightarrow 2\pi^- + 2\pi^+ + \pi^0$$

in quark terms:      uud + $\overline{uud}$ → 2$\overline{u}$d + 2u$\overline{d}$ + u$\overline{u}$ (or d$\overline{d}$)

u-quark number, $U$:    2 + (−2)   (−2) +   2   +   0 (see Pointer)

d-quark number, $D$:    1 + (−1)     2   + (−2) + 0

So $U$ and $D$ are both conserved in the interaction – they are both 0 before and after.

### » Pointer

It doesn't matter whether the $\pi^0$ is u$\overline{u}$ or d$\overline{d}$. In both cases, $U = D = 0$.

## (b) Electromagnetic interaction

- Particle decays have short lifetimes, typically ~$10^{-12} - 10^{-18}$ s.
- All the particles involved are charged or have charged components, e.g. interaction 1 above.
- The reaction is very likely to happen when particles collide, e.g. reaction 1 above.
- One or more photons may be emitted.
- There is no change in quark flavour.

### quickfire

⑦ The $\pi^0$ usually decays into two photons:

$$\pi^0 \rightarrow \gamma + \gamma$$

Explain what you would expect the lifetime of a $\pi^0$ to be.

## (c) Weak interaction

- Particle decays have long lifetimes, typically ~$10^{-10}$ s and longer.
- Neutral leptons (neutrinos) are involved.
- There may be a change of quark flavour, e.g. reaction 3 above.
- The reaction is very unlikely to happen when two particles collide, e.g. $p + p \rightarrow {}^2_1H + e^+ + \nu_e$.

### » Pointer

The weak interaction example
$p + p \rightarrow {}^2_1H + e^+ + \nu_e$
is the first stage in the proton–proton chain in the Sun. The lifetime of any proton in the Sun's core is ~$10^9$ years.

## (d) Which interaction governs a particular event?

The rule is that the strongest interaction which could possibly be responsible is the one that occurs.

For example, it is *possible* for the reaction $p + p \rightarrow p + n + \pi^+$ to proceed by way of the weak interaction, but it is overwhelmingly more likely for the strong force to be responsible.

### Grade boost

If a reaction can proceed via the strong interaction, it does so. If it cannot it proceeds via the e-m unless there is a change of quark flavour, in which case it is weak.

⑧ How can you tell, in terms of the 'tell-tale signs' that the weak interaction controls β– decay?

⑨ State the values of $U$ and $D$ for $\Delta^+$ and $p + \pi^0$. Comment.

⑩ The other common decay of a $\Delta^+$ produces a charged pion and another particle. Write this decay:
(a) At the particle level.
(b) At the quark level.

⑪ (a) Explain why the reaction
$$e^+ + e^- \rightarrow \pi^+ + \pi^-$$
cannot be controlled by the strong force.
(b) Explain why the reaction is controlled by the electromagnetic interaction.

# 1.7.8 Some particle interactions

## (a) β⁻ decay – a weak decay

A typical $\beta^-$ decay is: $^{14}_{6}C \rightarrow {}^{14}_{7}N + e^- + \overline{\nu_e}$.

We can write it as $n \rightarrow p + e^- + \overline{\nu_e}$ because it comes about when a neutron decays into a proton. We can also look at the decay at the quark level, remembering that n = udd and p = uud, so essentially the decay is

$$udd \rightarrow uud + e^- + \overline{\nu_e}$$

or simply $\qquad d \rightarrow u + e^- + \overline{\nu_e}$

The half-life of C-14 is ~5700 years. Charge, baryon number and lepton number are all conserved but quark flavour isn't. Using this information you should be able to show that the decay is controlled by the weak interaction.

## (b) Δ decay

The lifetime of the $\Delta$ particles is about $5 \times 10^{-24}$ s and one of the common decay modes of the $\Delta^+$ is:

$$\Delta^+ \rightarrow p + \pi^0$$

Which force controls this decay? Can you justify your answer?[2]

## (c) e-e⁺ annihilation

Low energy electron-positron annihilation usually results in the emission of two photons as in Fig. 1.7.2. For very high energy collisions, pions can be produced:

$$e^+ + e^- \rightarrow \pi^+ + \pi^-$$

You should be able to show that the conservation laws are all obeyed in this event, but which force controls it?

Hints:

- The U and D numbers are conserved (check this).
- There are leptons as well as quarks.
- All the particles are charged.

(See Quickfire 11)

---

[2] Strong interaction: very short lifetime, all hadrons, no change of quark flavour (see Quickfire 9).

# Extra questions

1. Calculate the total $U$ and $D$ numbers before and after the interaction
$$p + p \rightarrow p + n + \pi^+.$$

2. From what you know about the strong and e-m force, how do you know that neutrinos (and antineutrinos) can't experience either.

3. A $\pi^+$ can (occasionally) decay via: $\pi^+ \rightarrow e^+ + \nu_e$. Show that this decay satisfies the conservation rules and explain which interaction is responsible.

4. A $\pi^-$ can decay by a similar mode to that in question 3. Write the decay equation.

5. The $\rho^+$ meson decays via the strong interaction to a $\pi^+$ and another hadron. Write the equation for this decay and explain why this is the only possibility.

6. The following occurs in the proton–proton chain of reactions in the Sun:
$$^7_4\text{Be} + e^- \rightarrow \, ^7_3\text{Li} + Y$$
Write this reaction at the quark level and identify particle Y, explaining your answer.

7. The last stage of the ppI branch of the proton–proton chain in the Sun is
$$^3_2\text{He} + \, ^3_2\text{He} \rightarrow \, ^4_2\text{He} + 2^1_1\text{H}$$
where we have used the nuclear symbol for the protons.
Which of the interactions is responsible?

8. Solar neutrinos can be detected on Earth using dry-cleaning fluid! In the process, occasionally a $^{37}_{17}\text{Cl}$ nucleus is converted into $^{37}_{18}\text{Ar}$ with the emission of another particle. Write an equation for this reaction, identifying the emitted particle.

9. The solar neutrino flux (i.e. the number of neutrinos per m$^2$ per second) at the Earth is enormous: $\sim 7 \times 10^{14}$ m$^{-2}$ s$^{-1}$, yet the neutrino detector of volume 1000 m$^3$ only detects a few neutrinos per year. Suggest a reason for this.

10. Some (most) of the following 'reactions' cannot happen because they violate one or more of the conservation laws. Identify these. For the others, state which interaction controls it.
(i) $\bar{p} + \bar{p} \rightarrow \bar{n} + \bar{p} + \pi^0$, (ii) $e^+ + e^- \rightarrow n$, (iii) $p + p \rightarrow n + \overline{\Delta^-} + \pi^+$,
(iv) $\overline{\Delta^-} \rightarrow \pi^+ + \bar{n}$, (v) $u + e^- \rightarrow d + \nu_e$, (vi) $p + e^+ \rightarrow n + \overline{\Delta^-} + \nu_e$,
(vii) $\rho^- \rightarrow \pi^- + \pi^0$

# Knowledge and Understanding

## Component 2

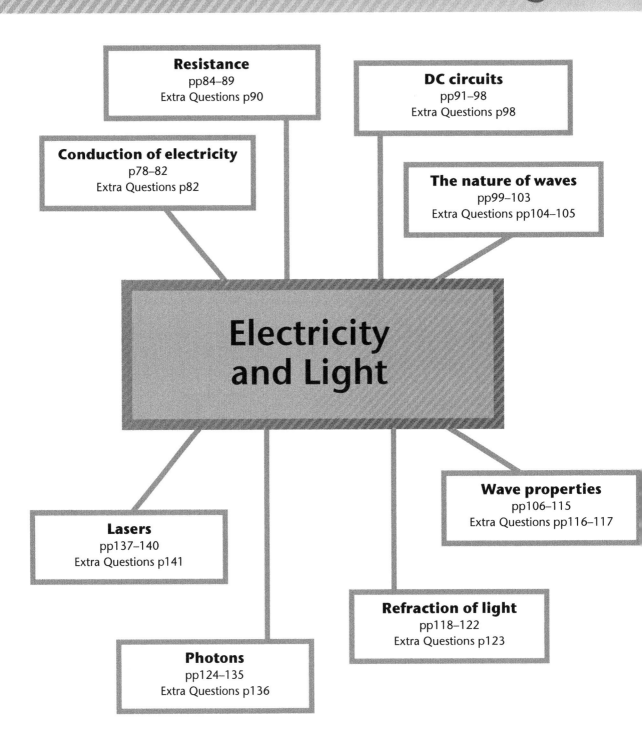

**Resistance**
pp84–89
Extra Questions p90

**DC circuits**
pp91–98
Extra Questions p98

**Conduction of electricity**
p78–82
Extra Questions p82

**The nature of waves**
pp99–103
Extra Questions pp104–105

# Electricity and Light

**Wave properties**
pp106–115
Extra Questions pp116–117

**Lasers**
pp137–140
Extra Questions p141

**Refraction of light**
pp118–122
Extra Questions p123

**Photons**
pp124–135
Extra Questions p136

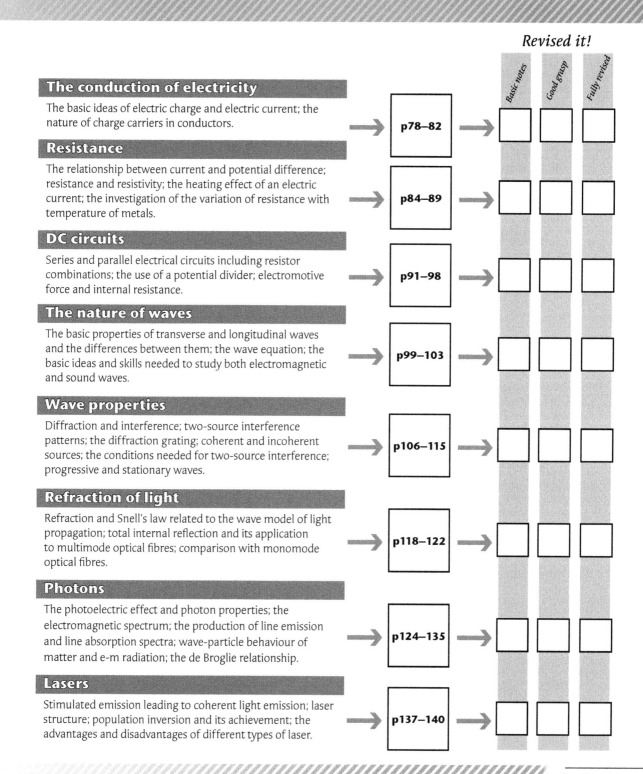

*Revised it!*

Basic notes  Good grasp  Fully revised

## The conduction of electricity

The basic ideas of electric charge and electric current; the nature of charge carriers in conductors.

p78–82

## Resistance

The relationship between current and potential difference; resistance and resistivity; the heating effect of an electric current; the investigation of the variation of resistance with temperature of metals.

p84–89

## DC circuits

Series and parallel electrical circuits including resistor combinations; the use of a potential divider; electromotive force and internal resistance.

p91–98

## The nature of waves

The basic properties of transverse and longitudinal waves and the differences between them; the wave equation; the basic ideas and skills needed to study both electromagnetic and sound waves.

p99–103

## Wave properties

Diffraction and interference; two-source interference patterns; the diffraction grating; coherent and incoherent sources; the conditions needed for two-source interference; progressive and stationary waves.

p106–115

## Refraction of light

Refraction and Snell's law related to the wave model of light propagation; total internal reflection and its application to multimode optical fibres; comparison with monomode optical fibres.

p118–122

## Photons

The photoelectric effect and photon properties; the electromagnetic spectrum; the production of line emission and line absorption spectra; wave-particle behaviour of matter and e-m radiation; the de Broglie relationship.

p124–135

## Lasers

Stimulated emission leading to coherent light emission; laser structure; population inversion and its achievement; the advantages and disadvantages of different types of laser.

p137–140

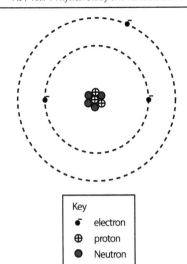

Key
- ● electron
- ⊕ proton
- ● Neutron

*Fig. 2.1.1 Atomic structure*

## quickfire

① Express the sizes of the atomic particles given in the main text in standard form.

## ≫ Pointer

The symbol for the magnitude of the charge on the electron is $e$, which has a value of $1.60 \times 10^{-19}$ C. The charge on the electron is $-e$. The charge on the proton is $e$.

## quickfire

② When an ebonite rod is rubbed with fur it gains $2.5 \times 10^{10}$ electrons from the fur. What is the resulting charge on the rod?

## ⚑ Grade boost

It's surprising how often you have to multiply or divide by $1.60 \times 10^{-19}$ ($e$) in A-level Physics. The only problem is deciding whether to divide or multiply. If in doubt do both and see which looks more likely!

# 2.1 The conduction of electricity

Before getting started with anything new, here are some very basic facts that you should already know:

- Protons (located in the nucleus of an atom) have a positive charge.
- Electrons (orbiting the nucleus) have a negative charge.
- Like charges repel, opposites attract.
- Electrical current is a flow of charge.
- Conductors allow charge to flow through them.
- Insulators do not allow charge to flow through them.

The 'solar system' representation of a $^7_3$Li atom in Fig. 2.1.1 is very much a schematic one. Electrons do not 'orbit' in the conventional sense and the atom is a sphere and not a plane. Also the particle sizes are misleading: the diameter of an atom ~0.1 nm, the nucleus ~1 fm and the electron has no measurable size (it is often given as 1 am).

## 2.1.1 Electrical charge

Charge is expressed in units of coulomb (C) and can be measured with a device called a coulomb meter. Some coulomb meters measure static charge (see Fig. 2.1.2) while others sit in a circuit like an ammeter and tell you the total amount of charge that has passed through.

The coulomb is rather a large unit of charge and you get some idea of this when you know that the charge on an electron is $-1.60 \times 10^{-19}$ C. The symbol for charge is $Q$.

*Fig. 2.1.2 Coulomb meter*

### Example

A glass rod is rubbed with a silk cloth and, as a result, it acquires a charge of +25.0 nC. Explain this in terms of electrons.

### Answer

Rubbing with the silk cloth causes electrons to be transferred from the glass rod to the silk cloth, leaving the rod with fewer electrons than protons and hence a net positive charge.

We calculate the number of electrons as follows:

$$\text{Number of electrons} = \frac{Q}{e} = \frac{25.0 \times 10^{-9}}{1.60 \times 10^{-19}} = 1.56 \times 10^{11}$$

# 2.1.2 Current and charge

You already know that **current** is a flow of charge but you might not have come across the more precise definition. Current is the rate of flow of charge. This appears as an equation on the Data Booklet:

$\dfrac{\Delta Q}{\Delta t}$ but you can usually write $\dfrac{Q}{t}$ without losing marks!

This equation means that if you look at any point in a circuit and a charge $Q$ passes that point in time $t$, then the current is $\dfrac{Q}{t}$.

The unit of current is the ampère (A) but from the equation $A = C\ s^{-1}$ (i.e. a current of one amp is equal to 1 coulomb of charge passing per second).

## Example

If 25 C of charge passes through a point in a circuit in 1.0 minute, what is the current?

## Answer

$I = \dfrac{\Delta Q}{\Delta t} = \dfrac{25}{60} = 0.40\ A$      (note the conversion 1 minute = 60 s)

## A slightly trickier example

The current in an LED is 35.2 mA. How many electrons flow through it in 1 hour?

[It's not really so very nasty, but there are two stages, a formula rearrangement and a unit conversion to it.]

## Answer

$I = \dfrac{\Delta Q}{\Delta t} \quad \rightarrow \quad \Delta Q = I\Delta t = 35.2 \times 10^{-3} \times 3600 = 127\ C$

$\therefore$ Number of electrons $= \dfrac{127}{1.60 \times 10^{-19}} = 7.9 \times 10^{20}$

**Key Term**

Electric **current**, $I$, is the rate of flow of charge, $Q$.

**quickfire**

③ The total amount of charge that passed a point in a circuit was 18.6 C and the current was a constant 450 mA. For how much time was the charge flowing?

**» Pointer**

The LED question is a bit trickier because it's a two-step question. You need to be thinking ahead, e.g. I can get the charge from current and time then I can get the number of electrons by dividing by $1.60 \times 10^{-19}$ C.

**quickfire**

④ The total number of electrons that pass through a battery in 1 hour is $7.87 \times 10^{23}$. Calculate the mean current.

**» Pointer**

The two equations
$$v = \frac{\Delta x}{\Delta t} \quad \text{and} \quad I = \frac{\Delta Q}{\Delta t}$$
have the same structure, with the analogues:
$$v \leftrightarrow I \text{ and } \Delta x \leftrightarrow \Delta Q$$
Hence, if the gradient of an $x$–$t$ graph is $v$, the gradient of a $Q$–$t$ graph is $I$.

# 2.1.3 Calculations involving varying currents

In Section 1.2 we saw that we can use the gradients of $x-t$ and $v-t$ graphs and the area under a $v-t$ graph to calculate velocities, accelerations and displacements when these quantities vary. In the same way we can make use of the following properties of charge and current graphs:

- The gradient of a $Q-t$ graph is the current (see Pointer).
- The area under an $I-t$ graph is the total charge flow.

### Example

During the discharge of a school Van der Graaf generator, the variation of current with time is as shown in Fig. 2.1.3. Estimate the total charge which was originally stored.

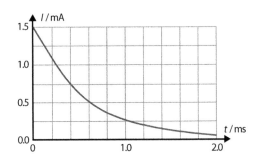

Fig. 2.1.3 Van der Graaf generator discharge

### Answer

Each square on the grid represents = 0.05 µC.

Using the square-counting method: There are 18 squares below the line

(counting $> \frac{1}{2}$ square as a full square and $< \frac{1}{2}$ square as 0), so

$$Q = 18 \times 0.05 \ \mu C = 0.90 \ \mu C$$

## ≽≫ A fussy point

We all fall into the trap occasionally but ....try to avoid referring to a 'flow of current'. Charge (like water) can flow. Current is defined as the 'rate of flow of charge' and it doesn't make any sense to talk about a 'flow of the rate of flow' of anything!

## ≫ Pointer

The delocalised electrons collide with each other and the lattice ions, a process causing energy transfer. The free movement and energy transfer accounts for the high *thermal conductivity* of metals. You will need to remember the free-electron model when we explain the variation of resistance with temperature.

### Key Terms

The **drift velocity** is the small mean velocity of electrons in a conductor due to an applied pd.

In $I = nAve$, the symbol $n$ represents the number of free electrons per unit volume (i.e. per m³).

# 2.1.4 Conduction in metals

### Bonding in metals

As part of your GCSE Chemistry course you learnt that atoms in metals are bonded by losing one or more electrons, so that they become positive ions. These 'lost' electrons are often called *delocalised electrons* [or 'free electrons'] because they are free to move throughout the metal rather than being located in one atom. The positive ions are held together by the negatively charged mobile electrons. In the work that follows it is important to distinguish between these mobile electrons, the movement of which constitutes the current and the majority of the electrons which are in inner orbits: they ain't going nowhere and therefore cannot contribute to the electric current.

The motion of the free electrons is random and similar to the motion of gas particles, i.e. they move about quickly then have collisions. Because of this, sometimes the free electrons in a metal are called a 'free electron gas'. Electrons usually travel about 40 nm at a speed of around $2 \times 10^6$ m s$^{-1}$ in between collisions, which means that their time between collisions is about 20 fs. Before a potential difference (pd or voltage) is applied, the mean velocity of these electrons is exactly zero – their motion is completely random in all directions so the vector average of their velocities is zero. As soon as a pd is applied these electrons will be accelerated by the pd in between collisions and they'll end up with a tiny little average velocity that will constitute a current.

Consider the following cross-section of a wire where we have a lot of electrons all moving to the right with a drift velocity $v$.

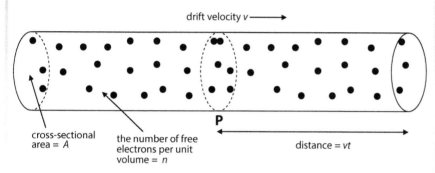

Fig. 2.1.4 Electron drift in a wire

Each of those dots represents a free electron. In order to calculate the current you need to know the charge passing a certain point every second – let's look at point P. You need to consider some interval of time $t$. In this time, the electrons have moved on a distance $vt$. So, if you can count all the electrons in the tube to the right of point P you can calculate the current.

These are the four essential steps:

1  Volume (right of P) = $A \times vt$     (csa × length of the cylinder)
2  Number of electrons = $nAvt$     (number = $n$ × volume)
3  Charge = $nAvte$     (charge = number × electronic charge)
4  Current, $I = \dfrac{nAvte}{t} = nAve$     (current = charge / time)

i.e. $I = nAvE$          QED

Remember: to derive $I = nAvE$, you need no more than a clear diagram and the four steps above.

## Example

In copper there are $8.5 \times 10^{28}$ free electrons per cubic metre (which means that $n = 8.5 \times 10^{28}$ m$^{-3}$). A copper wire of diameter 0.213 mm carries a current of 0.35 A. Calculate the drift velocity of the electrons.

## Grade boost

It is important to remember the diagram and the four steps. This proof is a favourite of examiners: there aren't all that many proofs they can ask you!

## Grade boost

Sometimes you are asked to state the meaning of $n$. A common answer given is, 'the number of electrons per m$^3$'. Wrong! It's the number of *free* electrons [or *delocalised* electrons] per m$^3$.

## quickꞇire

⑤ The current in a tungsten wire is 1.20 A and the drift velocity of electrons is 2.35 mm s$^{-1}$. The number of free electrons per m$^3$ in tungsten is $6.3 \times 10^{28}$ m$^{-3}$. Use the table to find the gauge of the wire.

| Standard Wire Gauge | Diameter / mm |
|---|---|
| 31 | 0.295 |
| 32 | 0.274 |
| 33 | 0.254 |
| 34 | 0.234 |
| 35 | 0.213 |
| 36 | 0.193 |
| 37 | 0.173 |
| 38 | 0.152 |

**» Pointer**

In order to provide the same current, electrons have to move faster in the thinner wire. This is worth remembering.

**Answer**

$$A = \pi r^2 = \pi \frac{d^2}{4} = \pi \frac{(0.213 \times 10^{-3})^2}{4} = 35.6 \times 10^{-9}\ \text{m}^2$$

then using $I = nAve$

$$\rightarrow \quad v = \frac{I}{nAe} = \frac{0.35}{8.5 \times 10^{28} \times 35.6 \times 10^{-9} \times 1.60 \times 10^{-19}} = 7.2 \times 10^{-4}\ \text{m s}^{-1}$$

There are two important things to remember about the figures here:

1. $n$, the number of free electrons per $m^3$, is a very large number for metals, around $10^{28}$ free electrons per $m^3$.
2. $v$, the drift velocity, is a small number for metals, less than a mm per second even for this thin wire.

## Extra questions

1. The electric charge on a comb after combing a person's hair is –25 pC. Explain this in terms of electrons.

2. The capacity of a battery is often expressed in ampère hours (A h).
   (a) Explain why A h is a unit of charge.
   (b) A battery has a stated capacity of 500 mA h. Express this in coulomb.

3. Calculate:
   (a) The net charge on a $^{235}_{92}U$ atom.
   (b) The net charge on an $Al^{3+}$ ion.
   (c) The total charge on all the electrons in 1 mol of $^{12}_{6}C$.
   ($N_A = 6.02 \times 10^{23}\ \text{mol}^{-1}$)

4. A research Van der Graaf generator is charged. During this process, the charge stored on it varies with time as shown in the graph. As accurately as you can, draw a graph of the charging current against time.

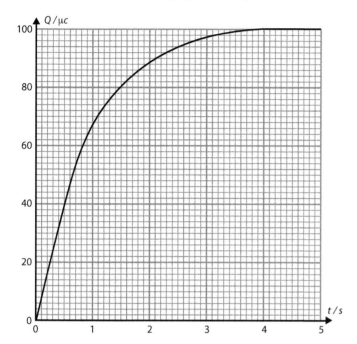

5. Two conducting wires, **A** and **B**, are made from the same material. The diameter of **A** is twice that of **B**. The length of **A** is three times that of **B**. Wire **B** carries a current, twice that of **A**. Compare:

(a) The number of free electrons in **A** and **B**.

(b) The free electron concentration in **A** and **B**.

(c) The drift velocity of the electrons in **A** and **B**.

## Key Terms

The **potential difference** (pd), $V$, between two points is the energy converted from electrical potential energy to some other form per unit charge flowing from one point to the other.

Unit: volt, $V = J\,C^{-1}$

The **resistance**, $R$, of a conductor is defined by the equation:

$$R = \frac{V}{I}$$

where $V$ is the pd across it and $I$ the current through it.

Unit: ohm, $\Omega = V\,A^{-1}$

## ⟫ *Pointer*

Pd is made even more confusing by the fact that the symbol ($V$) and the unit (V) are so similar – and in handwriting they are identical. Look out for possible proof-reading errors by the author.

### quickpire

① Calculate the energy transferred when a charge of 28 C flows through a pd of 12 V.

### quickpire

② The current in an 82 Ω resistor is 72 mA. Calculate:
(a) The pd across the resistor.
(b) The charge that flows in 1 min 20 seconds.
(c) The energy dissipated in the resistor in this time.

# 2.2 Resistance

Although it is just called resistance, this topic deals with a wide range of properties of conductors and electric current, including energy transfer. We start with **potential difference** – usually abbreviated to pd, and often referred to as voltage – which is perhaps the most difficult concept in AS Physics (but don't let that put you off).

## 2.2.1 Potential difference (pd) and resistance

The actual definition of pd you can see in the Key terms and this comes up regularly on physics papers. The definition itself might not make all that much sense to you and this is where the penguins come to the rescue.

*Fig.2.2.1 Playful penguins*

If you look at the toy in Fig. 2.2.1 you'll see that it works through having a mechanised elevator lift the penguins to the top, the penguins then slide to the bottom where they get picked up again by the elevator. This is very similar to the way charges flow around a circuit.

The elevator is similar to a cell, it provides the gravitational potential energy (GPE) to the penguins who then gradually lose all this GPE until they arrive back at the elevator which provides the GPE again.

In an electrical circuit, the cell provides electrical potential energy (EPE) to the charges which then lose this EPE gradually until they arrive back at the cell where they once more gain EPE.

The reason why this comparison explains that pd is associated with **energy per unit charge** is this:

- If the elevator were empty it wouldn't be doing any work (the elevator is extremely light and frictionless!).
- The more penguins that are on the elevator the more work is done by the elevator. In fact, the work done by the elevator is proportional to the number of penguins that have been lifted by it.

In the same way, the work done by a cell in a circuit is proportional to the quantity of charge that flows through it. The work done ($W$) by a cell of emf ($V$) when a charge ($Q$) flows is $W = QV$. Rearranged this gives:

$V = \dfrac{W}{Q}$, which is the definition of pd applied to the whole circuit.

The **resistance** of a conductor is defined by the equation $R = \dfrac{V}{I}$.

# 2.2.2 I–V graphs and Ohm's law

There are only two **characteristic** graphs that you need to know. You also need to be able to describe how such graphs are obtained experimentally – see Section 2.2.6.

## (a) Metallic conductor at a constant temperature

The graph is a straight line through the origin (signifying $I \propto V$). The reason why the pd is usually (but not always) plotted on the $x$-axis is that the pd is the variable that you usually change and the current is the variable that you subsequently measure. Another thing that you often do on an $I$–$V$ graph is to plot the negative values as well as the positive values; this is because not all electrical devices behave the same when the voltage is reversed, e.g. a diode.

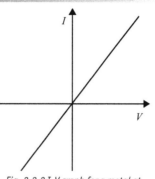

*Fig. 2.2.2 I–V graph for a metal at constant temp*

## (b) Lamp filament

There are two important parts to this graph that you need to know:

- For low voltages the graph is straight, i.e. $I \propto V$.
- The gradient decreases smoothly as the pd increases. This because the resistance is increasing.

Why does the resistance increase? Because the temperature of the filament is increasing.

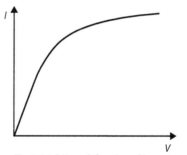

*Fig. 2.2.3 I–V graph for a lamp filament*

At its full rating, the tungsten filament usually operates at about 2500°C.

## (c) Ohm's law

Metals at a constant temperature are examples of what are called **ohmic conductors**, i.e. **Ohm's law** applies to them. Ohm's law is an experimental relationship. To understand why Ohm's law does not apply to lamp filaments we need to return to the 'free electron gas' model (see Section 2.1.4) and examine the origin of resistance.

When a pd is applied across a conductor, it exerts a force on each free electron, which is accelerated (towards the positive of the pd). This additional non-random velocity lasts until the next collision which randomises the electron's velocity once more. These additional velocities are all in the same

>> **Pointer**

Even though these two graphs are the only ones you need to know, the examiner could give you any characteristic and then ask questions about it.

### quickfire

 A mains (230 V) electric kettle element dissipates a power of 3.0 kW. Calculate:
(a) The current.
(b) The resistance of the kettle element.

### Grade boost

Students often state that the gradient of an $I$–$V$ characteristic is $1/R$. This is only true for ohmic devices. For any other graphs, e.g. in Fig. 2.2.3, $R$ must be calculated using $V/I$.

direction so, on the average, the electrons 'drift' towards the positive; this drift constitutes the current.

There are two important things that happen as the temperature rises:

- The electrons travel faster between collisions.

- The metal lattice ions vibrate more (i.e. with a greater energy).

Both these effects decrease the time between collisions. If the collisions occur more frequently, then the electrons will have a smaller drift velocity because they can't be accelerated as much between collisions ($\Delta v = a\Delta t$). This leads to a decreased current, for the same voltage and therefore a higher resistance.

## Grade boost

Only equation [1], $P = IV$, is in the Data Booklet. You need to learn the other two or learn how to derive them. Even better, learn all three **and** how to derive them!

### quicϰꜰire

④ A mains (230 V) hairdrier has a power rating of 1.2 kW. Calculate the resistance of its element. [Do this in a single stage, rather than calculating $I$ then $R$.]

## 2.2.3 Electrical power

If we combine the definition of pd with the definitions of power and current, you can derive three very useful equations for electrical power:

Start first with the work done $\hspace{3cm} W = QV$

Then divide by the time taken $\hspace{2cm} \dfrac{W}{t} = \dfrac{QV}{t}$

But $\dfrac{W}{t}$ = power, $P$ and $\dfrac{Q}{t}$ = current, $I$ $\hspace{1cm} \therefore \hspace{1cm} P = IV \hspace{0.5cm}$ [1]

But $V = IR$, so substituting in [1] for $V$ and simplifying $\;\rightarrow\; P = I^2 R \hspace{0.5cm}$ [2]

Or substituting in [1] or [2] for $I$ and simplifying $\;\rightarrow\; P = \dfrac{V^2}{R} \hspace{0.5cm}$ [3]

## 2.2.4 The resistance of an electrical conductor

You need to remember how the resistance of a conductor depends upon its composition, its dimensions and its temperature. You also need to describe how to investigate these relationships experimentally. See also Section 2.2.6.

### (a) Resistivity

Whereas resistance, $R$, is a property of a particular piece of material (e.g. a resistor), resistivity, $r$, is a property of a material.

For wires made out of the same material, the resistance is:

- directly proportional to the length, $l$ $\hspace{3cm} R \propto l$

- inversely proportional to the cross-sectional area, $A$ $\hspace{0.5cm} R \propto \dfrac{1}{A}$

The resistance of the wires also depends upon the material of the wire. We define the resistivity, $\rho$, by the equation:

$$R = \frac{\rho l}{A} \text{ (see Grade boost)}$$

## Grade boost

When you define a quantity by an equation, it is important to define all the terms in the equation, i.e.
$l$ = length of the conductor
$A$ = cross-sectional area of the conductor
$R$ = resistance of the conductor.

To work out the units of $\rho$ we'll make it the subject of the equation:

$$\rho = \frac{RA}{l}.$$

Then the unit of $\rho$, $[\rho] = \dfrac{\Omega\ \text{m}^2}{\text{m}} = \Omega\ \text{m}$

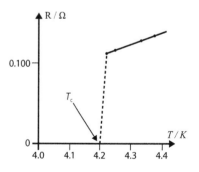

## quickfire

⑤ Rearrange
$R = \dfrac{\rho l}{A}$ to make
(a) $l$ and (b) $A$ the subject.

## (b) Variation of resistance with temperature

As the characteristic of a filament lamp shows, the resistance of a metal increases with temperature. For school laboratory temperature ranges (usually between 0 °C and 100 °C) the variation of resistance with temperature is almost linear. In fact, for pure metals, this linear behaviour extends down to temperatures lower than −200 °C. The graph in Fig. 2.2.4 shows this for pure platinum.

### ≫ Pointer

The variation of the resistance of a metal sample with temperature is almost entirely due to changes in the resistivity. The length and cross-sectional area also change, but this effect is negligible.

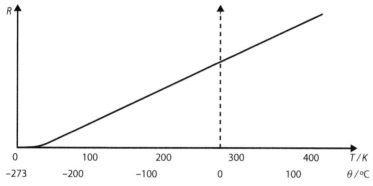

Fig. 2.2.4 Platinum R–T graph

## quickfire

⑥ The resistance of a copper wire at 20 °C is 10.3 Ω. At 100 °C its resistance is 13.4 Ω. Estimate its resistance at 200 °C. State your assumption.

# 2.2.5 Superconductivity

On 8 April 1911, the Dutch physicist Heike Kamerlingh Onnes found that the resistance of a solid mercury wire at 4.2K (−269 °C) suddenly dropped to zero, as shown in Fig. 2.2.5. This was an incredible discovery that started a whole new research area into **superconductors** and Heike Kamerlingh Onnes was later rewarded with a Nobel Prize.

Many but not all metals are superconductors at temperatures close to absolute zero (i.e. at temperatures of a few kelvin or around −270 °C). Some examples of superconductors are aluminium, tin, lead and mercury. As they are cooled through a special temperature called the **superconducting transition temperature,** $T_c$, their resistance drops suddenly to zero.

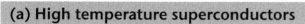

Fig. 2.2.5 Superconducting transition in mercury

## (a) High temperature superconductors

The superconducting transition temperature for pure metals is always within a few kelvin of absolute zero (below ~30 K). Since 1986 several classes of ceramic materials have been discovered which have a transition temperature above the boiling point of liquid nitrogen. These are called **high-temperature superconductors (HTS)**.

## (b) Using superconductors

The giant magnets at the CERN particle accelerators are superconducting. They are cooled close to absolute zero by liquid helium. Because their resistance is zero they dissipate no energy when carrying a current (if $R = 0$, then $P = I^2R = 0$ also), so:

- There is no need to design systems for heat to be conducted away.
- The energy costs are kept low.

For many medical and industrial applications, the engineering challenges of working at such low temperatures, to say nothing of the cost, are too great. In these applications, such as MRI scanners in hospitals, high temperature superconductors are used.

# 2.2.6 Specified practical work

## (a) I–V characteristics

The simplest circuit to draw for this practical is shown in Fig. 2.2.6, where the arrow through the cell symbol, represents a variable voltage supply.

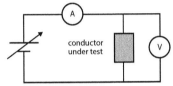

Fig. 2.2.6 Simple I–V circuit

The two other possible circuits in Fig. 2.2.7 use a fixed voltage supply and (a) a variable resistor or (b) a potentiometer.

Whichever circuit is used, the procedure is to set the current (or pd) to a low value [by turning the variable power supply down, the variable resistor up or the potentiometer down] and reading the current and pd values. Then increase the current (or pd) in a series of steps to obtain at least five pairs of current /pd readings. In this experiment, there is little need for repeats apart from to pick up mistaken readings. Finally, plot a graph of current against pd.

When investigating a wire at a constant temperature, either the current should be kept low enough that minimal rise in temperature occurs or inserting a push switch so that the current is switched on for a minimal length of time. Alternatively, the (insulated) wire should be immersed in a water bath to maintain a constant temperature.

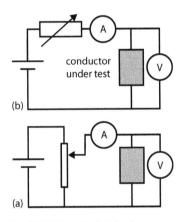

Fig. 2.2.7 Alternative I-V circuits

## (b) Determination of the resistivity of the material of a wire

The easiest method is to tape a length (> 1 m) of the uninsulated wire to a metre rule and connect it to a multimeter, set to the lowest resistance range using the meter leads and crocodile clips – see Fig. 2.2.8.

Obtain a series of readings of resistance, $R$, for a series of at least five different lengths, $l$, of wire up using the mm scale on the metre rule. Clip the crocodile clips together and take a reading to obtain the resistance of the leads (usually about 0.5 $\Omega$) and subtract this from each of the readings.

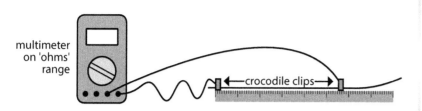

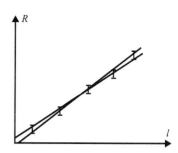

Fig 2.2.8 R-l graph with error bars

Fig. 2.2.9 Resistivity of a wire

From the results, a graph of $R$ against $l$ is plotted and the gradient, $m$, of the best-fit line measured. A micrometer or digital calliper is used to measure the diameter, $d$, of the wire at several different points and the mean value used to calculate the cross-sectional area $A$, using $A = \pi \dfrac{d^2}{4}$.

**Analysis**: Comparing $R = \dfrac{\rho l}{A}$, with $y = mx + c$, the gradient of the graph is $\dfrac{\rho}{A}$ and the intercept should be 0. So, rearranging: $\rho = mA$.

If we draw the error bars, as in Fig. 2.2.8, we can determine the uncertainty in $m$. This, together with the uncertainty in $A$, lets us work out the uncertainty in $\rho$.

**》 Pointer**

Fig. 2.2.8 shows the results of the $R$–$l$ experiment with error bars. To find the uncertainty in the value of $l$, the max/min gradients are measured.

## (c) The variation of resistance of a wire with temperature

The set up is as in Fig. 2.2.10. The ohm-meter is usually a multimeter set on the resistance range. It is best to use an insulated wire, e.g. an enamelled copper wire, because water is an electrical conductor, albeit not a very good one. The beaker is set up on a tripod and a bunsen burner used to provide the heating.

The zero reading on the ohm-meter is used as in the previous experiment and the apparatus is set up using cold water. The resistance and temperature are measured. The beaker is heated until the temperature has risen by about 10 °C, the bunsen removed, the water stirred (to obtain equilibrium) and the readings taken again. This is repeated up to the boiling point of water (100 °C). A graph is drawn of resistance (corrected for the resistance of the leads) against temperature.

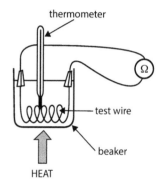

Fig. 2.2.10 R-θ variation

**》 Pointer**

To extend the range, an ice-water mixture can be used giving an initial temperature of 0 °C.

# Extra questions

1. Calculate:
   (a) The power of a 12 V fish tank heater which takes a current of 2.5 A.
   (b) The current taken by a 230 V, 15 W compact fluorescent lamp.
   (c) The resistance of an LED operating at 2.1 V, 19 mA.
   (d) The power of the radiation emitted by the 35% efficient LED in part (c).
   (e) The operating pd of a 1.5 kW heater with a resistance of 10.3 $\Omega$.

2. A travel kettle, designed for use in Britain, is labelled 230 V, 900 W. It takes 3 minutes and 20 seconds to bring 500 cm$^3$ of water to the boil.
   (a) Calculate the resistance of the element.
   (b) It is taken to the USA, where the mains pd is about half the British value, a different plug fitted and it is used to boil 500 cm$^3$ of water. Estimate the time it takes, stating any assumption you make.

3. The resistance of an 80.0 cm length of wire of diameter 0.305 mm is 15.8 $\Omega$. Calculate the resistivity of the material of the wire.

4. The material of wire **A** has twice the resistivity of wire **B**. It is also twice as long and has twice the diameter.
   (a) Compare the resistances of wires **A** and **B.**
   (b) The electron concentrations in wires **A** and **B** are equal. The two wires are connected across the same pd. Compare the electron drift velocities in the two wires.

5. A tungsten filament lamp which operates at 2500 °C is marked 240 V, 60 W. When connected across a 3.0 V battery at room temperature it takes a current of 60 mA. Estimate the pd needed to produce the same current at 500 °C.

# 2.3 DC circuits

This section of the book applies the ideas of the previous two sections and the principles of conservation of energy and charge to *simple* DC circuits. In this case, *simple* means having just one electrical supply – some of the resistor networks can be quite complicated. The major requirements for success in electrical circuit problems are to establish a systematic approach and to communicate clearly.

## 2.3.1 Conservation of charge

This can be applied to all circuits and essentially comes down to this – electrons are not created or destroyed in a circuit.

In addition, because they repel one another, electrons do not accumulate anywhere in the circuit (see Pointer).

This enables us to dispel the misconception that some young pupils have that the current somehow gets less as it goes around the circuit. In the circuit in Fig. 2.3.1, $I_1 = I_2$ because electrons don't suddenly disappear (or multiply or bunch up) in the resistor. If $2 \times 10^{16}$ free electrons enter the resistor every second, then $2 \times 10^{16}$ free electrons also exit every second, i.e. $Q_{in}$ per second = $Q_{out}$ per second.

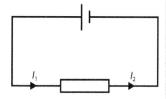

Fig. 2.3.1 $I_1 = I_2$

**Conclusion:** The current is the same at all points in a **series circuit** (even in the middle of the power supply).

We can apply this idea to parallel circuits:

Look at junction **A** in Fig. 2.3.2, where the current $I_1$ splits up into $I_2$ and $I_3$. You can't gain or lose any electrons at the junction (either **A** or **B**) and the result of this is:

$$I_1 = I_2 + I_3$$

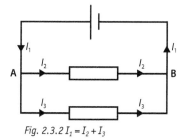

Fig. 2.3.2 $I_1 = I_2 + I_3$

**Conclusion:** The sum of the currents going into a junction is equal to the sum of the currents leaving the junction.

### Example

Find the unknown currents $w$, $x$, $y$ and $z$.

### Answer

We could just write the answer but we'll try and be systematic and communicative!

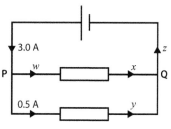

Fig 2.3.3 Currents problem

> ## ≫ Pointer
> As we saw in Section 2.2.4, metal wires possess electrical resistance (unless they are superconducting). However, the resistance of the connecting wires in electrical circuits in AS/A level Physics is generally much less than that of the circuit components and is taken to be negligibly small (i.e. zero). Hence the energy transfer in the connecting wires is taken to be zero.

> ## Key Terms
> Components are connected **in series** if they are joined by a single conductor which has no other conductors connected to it.
>
> A **series circuit** is one in which all the components are in series.
>
> Components are connected **in parallel** if both ends of the two components are joined by single conductors only.

> ## ≫ Pointer
> These current rules are also known as Kirchhoff's 1st law (or Kirchhoff's junction rule).

> ## quickfire
> ① In the example, show that current $z = 3.0$ A by a different method. [Hint: consider the power supply.]

**quickfire**

② In the circuit in Fig. 2.3.4, $I = 0.5$ A and $V_1 = 1.2$ V. Calculate:

(a) The energy transferred by the power supply to the circuit in 10 s.

(b) The resistance of the resistor.

(c) The pd across the lamp.

(d) The power dissipated in the lamp.

**》 Pointer**

The pds across all mains electrical appliances are the same (nominally 230 V in the UK). This is because they are connected in parallel. Adding another appliance doesn't change the pd across any other; switching an appliance off doesn't affect another appliance.

**quickfire**

③ A 2.5 kW electric kettle and a 2.0 kW convector heater are connected to a 230 V domestic ring-main. Calculate the total current when both devices are switched on.

At junction **P**: Current in = current out

$\therefore 3.0 = w + 0.5. \therefore w = 2.5$ A

Along **PQ**: Current into resistor = current out, $\therefore x = w = 2.5$ A

Similarly $y = 0.5$ A (current into and out of bottom resistor).

At junction **P**: Current out = current in, $\therefore z = 2.5 + 0.5 = 3.0$ A

## 2.3.2 Conservation of energy

The law of conservation of energy is a universal law, so we can apply it to electrical circuits, such as the one in Fig. 2.3.4. To do this we'll use the definition of pd: $W = QV$.

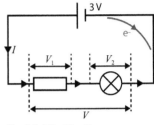
Fig. 2.3.4 $V = V_1 + V_2$

First we'll think about the resistor and lamp **series** pair and consider the story of an electron as it travels round the circuit.

It emerges from the battery's negative terminal (curved arrow) with some electrical potential energy. As it passes through the bulb filament it transfers $eV_2$, which the bulb radiates away. It then passes through the resistor where it transfers $eV_1$. If the pd across the two components is $V$ then the total energy transferred is $eV$.

$\therefore eV = eV_1 + eV_2$ and, dividing by $e$, $V = V_1 = V_2$

**Conclusion:** The total pd across components in series is the sum of the pds across the individual components.

Additionally, in Fig. 2.3.4, the energy transferred to the electron by the cell = $e \times 3$ V. The circuit cannot store electrical energy anywhere so, in this case, $V_1 + V_2 = 3$ V.

**Conclusion:** In a series circuit the sum of the pds across the components is the pd across the supply.

We can also apply conservation of energy to components in parallel. This time we'll think about *two* electrons which leave the negative terminal of the power supply and travel to the junction **A**. They meet up again at **B**, compare notes and continue on the power supply. They started off with the same electrical potential energy and end up having lost the same quantity.

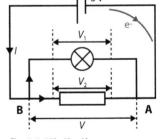
Fig. 2.3.5 $V = V_1 = V_2$

The wires have zero resistance, so the only place the electrons can have lost energy is in the lamp and the resistor. One has lost $eV_1$ and the other $eV_2$. So we conclude that $V_1 = V_2$. A voltmeter placed to measure $V$ would read the same because (again) the wires have zero resistance.

Additionally, by conservation of energy, each electron loses the 3 V $\times$ $e$ on its way around the circuit. $\therefore V = V_1 = V_2 = 3$ V

**Conclusion**: The pds across components in parallel are equal.

# 2.3.3 Combinations of resistances

## (a) In series

The combined resistance of components in series is the sum of their individual resistances. This is easily shown by applying voltage and current rules. The task is to replace the set of resistors in Fig. 2.3.6(a) with the single resistor in (b) so that the rest of the circuit (not shown) doesn't spot the switch!

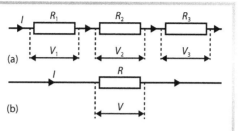

Fig. 2.3.6 Series combination of resistances

So the current, pd, power dissipated all remain the same. Luckily, if we sort out the current and pd, the power will look after itself (see Pointer).

For an undetectable switch, we must have $V = V_1 + V_2 + V_3$

The current must be unchanged, so $\dfrac{V}{I} = \dfrac{V_1}{I} + \dfrac{V_2}{I} + \dfrac{V_3}{I}$

But $R$ is defined by $R = \dfrac{V}{I}$, $\quad \therefore \quad R = R_1 + R_2 + R_3$

≫ **Pointer**

The pd rules are just energy conservation re-written for electrons!

## (b) In parallel

Once again the job is to find the single resistor that has the same effect in the circuit as the three shown.

This time we'll start with

$$I = I_1 + I_2 + I_3$$

Then

$$\frac{V}{R} = \frac{V}{R_1} + \frac{V}{R_2} + \frac{V}{R_3}$$

$\therefore$

$$\frac{1}{R} = \frac{1}{R_1} + \frac{1}{R_2} + \frac{1}{R_3} + \dots$$

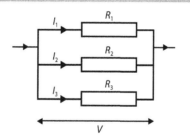

Fig. 2.3.7 Parallel combination of resistances

We've been a little naughty here, but the '+...' indicates that we can go on adding resistors in parallel and the equation just extends in the same way.

### Example

Calculate the combined resistance between **A** and **B**.

### Answer

First we calculate the resistance of the parallel combination:

$$\frac{1}{R} = \frac{1}{10} + \frac{1}{15}, \rightarrow R = 6\ \Omega. \text{ This}$$

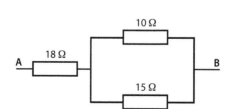

Fig. 2.3.8 Resistance network

$6\ \Omega$ is in series with the $18\ \Omega$ resistor.

$\therefore$ Combined resistance = $18\ \Omega + 6\ \Omega = 24\ \Omega$.

≫ **Pointer**

The combined resistance of two equal resistances in parallel is half that of the individual resistances. For three equal resistors it is one third; for $n$ equal resistances it is $1/n$.
Example: The effective resistance of five $100\ \Omega$ resistors in parallel is $20\ \Omega$.

**quickpire**

④ Calculate the effective resistance of a $22\ \Omega$ and a $33\ \Omega$ resistor connected (a) in series and (b) in parallel.

▲ *Grade boost*

When using $\dfrac{1}{R} = \dfrac{1}{R_1} + \dfrac{1}{R_2}$, don't forget to invert the value for $1/R$ otherwise your answer is likely to be nonsense.

## 2.3.4 The potential divider

### (a) Basic circuit

If there is no current in the $V_{OUT}$ leads the current, $I$, is the same in the two resistors

$\therefore \qquad V_{IN} = IR_1 + IR_2 \qquad [1]$

and $\qquad V_{OUT} = IR_2 \qquad [2]$

Dividing [2] by [1] and cancelling by $I$

$\rightarrow \qquad \dfrac{V_{OUT}}{V_{IN}} = \dfrac{R_2}{R_1 + R_2}$

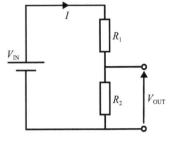

Fig. 2.3.9 Potential divider

The *potentiometer* is a potential divider made of a single conductor with a sliding contact. The output, $V_{OUT}$, in Fig. 2.3.9 is given by

$$V_{OUT} = V_{IN} \frac{l}{l_0}$$

The potentiometer circuit is commonly used in volume controls in audio amplifiers.

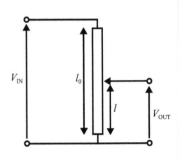

Fig. 2.3.10 Potentiometer

### Example

Calculate $V_{OUT}$ in Fig. 2.3.11 when the switch

(a) is open (as shown) and

(b) is closed.

### Answer

(a) The 82 Ω resistor is not part of the circuit, so:
$$V_{OUT} = \frac{27}{56 + 27} \times 12 = 3.9 \text{ V}$$

(b) $V_{OUT}$ is across the parallel combination of 27 Ω and 82 Ω.

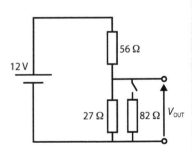

Fig. 2.3.11 Loaded potential divider

The resistance, $R$, of the parallel combination (see Grade boost) is:
$$R = \frac{27 \times 82}{27 + 82} = 20.3 \ \Omega$$
$$\therefore V_{OUT} = \frac{20.3}{56 + 20.3} \times 12 = 3.2 \text{ V}.$$

## (b) Potential divider sensing circuits

The potential divider used with an LDR or thermistor is the basis of light and temperature sensing circuits. You need to be able to draw and interpret these circuits. The resistance of the LDR decreases with increased light levels. [The thermistor's resistance decreases with temperature.] If one of these components is placed in the upper position in the circuit, as in Fig. 2.3.12, $V_{OUT}$ varies with light level / temperature; e.g. for an LDR:

$$V_{OUT} = \frac{R}{R_L + R} V_{IN} \text{, where } R_L \text{ is the resistance of the LDR.}$$

As the light level increases, $R_L$ decreases, so $V_{OUT}$ increases (see Pointer).

If a thermistor (or LDR) is placed in the lower position in the circuit, i.e. so $V_{OUT}$ is across it, then:

$$V_{OUT} = \frac{R_T}{R + R_T} V_{IN} \text{, where } R_T = \text{thermistor resistance}$$

You should be able to explain why, in this case, the output voltage decreases with temperature.

### Example

A frost alarm contains the temperature sensing circuit in Fig. 2.3.13. At 20 °C the thermistor temperature, $R_T$, is 200 Ω.

(a)  Calculate $V_{OUT}$ at 20 °C.

(b)  Explain why $V_{OUT}$ increases as the temperature falls.

### Answer

(a)  At 20 °C, $V_{OUT} = \frac{200}{1000 + 200} \times 5 = 0.83$ V.

(b)  As the temperature falls, the resistance of the thermistor rises, so $R_T$ is a larger fraction of the total resistance. The output is across $R_T$, so $V_{OUT}$ rises.

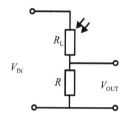

Fig. 2.3.12 Light-sensing circuit

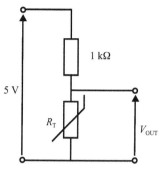

Fig. 2.3.13 Frost alarm

>> **Pointer**

If you find the argument in part (b) of the example difficult to follow, think about what $V_{OUT}$ would be if $R_T$ were (a) infinite (very low temperature) and (b) zero (very high temperature).

## 2.3.5 Tackling circuits systematically

The important things to remember when solving circuits in an exam are:

(a)  To have a plan.

(b)  To communicate what you are doing.

Communication is very important. It is all very well to write $V = IR$, but what are you applying it to – a single resistor, a combination, the whole circuit? We'll just have a look at a couple of examples – but you'll have to do most of the work!

### Example

Calculate the power dissipated in the 10 Ω resistor in Fig. 2.3.14.

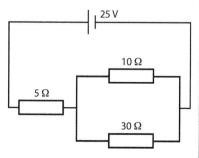

Fig. 2.3.14 Circuit problem

**quicKpire**

⑤ Look at the circuit in Fig.2.3.14 below.
 (i) Show that the resistance of the parallel combination is 7.5 Ω.
 (ii) Show that the pd across the 7.5 Ω combination is 15 V.
 (iii) Remember that the pd across a parallel combination is the same as the pd across each of the components in the combination. Hence calculate the power dissipated in the 10 Ω resistor.

## Grade boost

Remember to communicate, e.g.
1. To find the combined resistance of the parallel combination ....
2. The pd across the parallel combination is given by .....

## ≫ Pointer

The plan probably gives the neatest solution to the problem. But there are others. You've always got to do part 1 first. Then you could:

- Calculate the total resistance.
- Calculate the total current.
- Use the current to calculate the pd across the parallel pair......

You can probably think of other methods. Try them!

⑥ Try both suggested methods for the circuit puzzle to find the value of $R$.

### Key Term

The **emf** (electro-motive force) of a power supply is the energy converted from some form (chemical, in the case of a cell) to electrical potential energy per coulomb of charge passing through it.

## Grade boost

Students often say that emf is 'the pd across the cell with no current'. This is true but it is not the definition and will lose marks!

### Answer

First the plan:

1. Calculate the resistance of the parallel combination.
2. Calculate the pd across the parallel combination using potential divider theory.
3. Apply this pd to the 10 Ω resistor and calculate the power.

Next apply the plan: This is your job – see Quickfire 5.

The number of circuits an examiner can come up with is quite restricted but there are lots of ways of giving the information, which makes circuit questions rather like solving puzzles. Let's have a look at Fig. 2.3.15.

If the question is, 'What is the value of the resistor $R$ which is needed so that the lamp runs at its rating?' we need to come up with a cunning plan.

How about: calculate the current through the 240 Ω resistor, then use the total current and the known pd across the resistor to find its value?

How about: calculate the resistance of the lamp, then the combined resistance with the 240 Ω then use the potential divider formula for $R$?

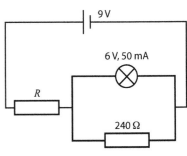

Fig. 2.3.15 Circuit puzzle

# 2.3.6 The emf and internal resistance of a power supply

The definition of the **emf**, $E$, of a power supply (its 'oomph') is a quantity which is regularly asked for in exam papers – so learn it.

The internal resistance, $r$, of a power supply is the resistance of its components (e.g. a cell's electrolyte, a dynamo's wires). Examiners often (but not always) include the internal resistance with diagrams of cells, in questions when they want you to take its effects into account – see Fig. 2.3.16

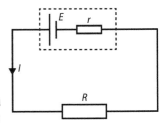

Fig. 2.3.16 E and r

The pd, $V$, across the terminals of a power supply is equal to the emf when the cell is not supplying any current. Usually it is less. We can see why by applying conservation of energy:

Power transferred from other forms into electrical = $EI$

Power delivered to the circuit = $VI$

Power wasted in the power supply = $I^2r$

∴ By conservation of energy:     $EI = VI + I^2r$

Dividing by $I$ and making $V$ the subject →     $V = E - Ir$

Sometimes an examiner likes to ask what $V = E - Ir$ means in terms of energy:

$E$  The energy converted in a cell from chemical to electrical potential per coulomb. [You could also say, 'The energy converted in the whole circuit per coulomb', because the cell is part of the circuit.]

$V$  The energy converted per coulomb outside the cell (or the energy dissipated in the resistor per coulomb).

$Ir$  The energy dissipated in the cell (or internal resistance) per coulomb.

### Example

A cell has an emf of 1.63 V and an internal resistance of 0.23 Ω. Calculate the current when the cell is connected to an 8.20 Ω resistor.

### Answer

$$I = \frac{E}{R + r} = \frac{1.63}{8.20 + 0.23} = \frac{1.63}{8.43} = 0.193 \text{ A (see Pointer)}$$

Note that the sorts of circuits in Section 2.3.5 could also include power supplies with internal resistance, so you need to be prepared for that.

## Circuits with multiple cells in series

The rules are simple:

- Internal resistances always add (resistances in series).
- The emfs add – unless one of the cells is the 'wrong way round', in which case it subtracts.

## 2.3.7 Specified practical work – the internal resistance of a cell

The simplest way of determining the internal resistance of a cell is to use the circuit in Fig. 2.3.17.

The method is:

- Close the switch and measure the current and pd using the ammeter and voltmeter.
- By adjusting the variable resistor, obtain a series of pairs of current, pd readings.
- Plot a graph of pd, $V$, against current, $I$ and measure the (negative) gradient.
- The internal resistance is minus the gradient, as shown by the equation of the graph:

$$V = E - Ir$$

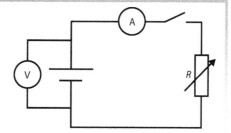

Fig. 2.3.17 Determination of internal resistance

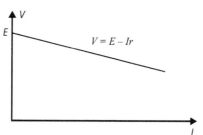

Fig. 2.3.18 V–I graph for a cell

**quicKpire**

⑦ An electrical cell converts 15 J of chemical energy every second and delivers a power of 12 W to a circuit with a current of 1.5 A.
Calculate:
(a) The power wasted inside the cell.
(b) The internal resistance.
Hint: $EI$, $VI$, $I^2 R$

**》 Pointer**

In Fig. 2.3.16 $V = IR$. Substituting for $V$ in $V = E - Ir$ gives $IR = E - Ir$ which can be rearranged to give
$$I = \frac{E}{R + r}.$$
This equation is not in the Data Booklet but it is very useful to learn. It lets you think: $E$ is the 'total voltage' in the circuit; the current is the 'total voltage divided by the total resistance'.

**quicKpire**

⑧ What are the emf and internal resistance of a battery of 6 cells, each of emf 1.50 V and internal resistance 0.05 Ω?

**quicKpire**

⑨ A voltage-current graph for a cell has intercept 1.65 V and a gradient of −2.5 Ω. State the emf and the internal resistance and calculate the maximum current which the cell can supply (i.e. when $R = 0$).

## Grade boost

In this experiment there is no point in taking repeat readings because the cell will lose a considerable amount of its stored chemical energy in supplying the large currents necessary to obtain good results.

**Note**: The switch in Fig 2.3.17 is important because it enables you to take current and pd readings quickly so that the cell does not become 'flat' before the end of the experiment.

## Minor variation in the method

If we use a set of resistors of known values for the variable resistor, we can do without either the ammeter or the voltmeter: we can calculate the pd from the resistance and the current (or the current from the resistance and pd) and then plot the graph.

## Extra questions

1. A student has a set of three resistors of values 4.7 Ω, 6.8 Ω and 8.2 Ω. By taking these singly or by connecting two or three of these in series and parallel what different values of resistance can the student achieve? [The author thinks there are 17 different values.]

2. An LED is to be used as an 'on indicator.' Calculate the series resistance that should be used if it is powered from a 5 V supply, so that the LED has a pd of 1.9 V across it and a current of 15 mA through it.

3. A 100 Ω resistor is connected in parallel with a series combination of a 47 Ω and a 33 Ω resistor and the network connected to a power supply. The pd across the 33 Ω resistor is 13.2 V. Calculate the pd across the power supply and the current through it.

4. The variation of the resistance of a thermistor with temperature is investigated and the following results obtained:

| Temperature / °C | −20 | −10 | 0 | 10 | 20 | 30 | 40 |
|---|---|---|---|---|---|---|---|
| Resistance / kΩ | 97 | 55 | 32 | 20 | 12 | 8 | 5 |

It is used in a potential divider circuit with a 10 kΩ resistor to make a temperature sensor, with an input pd of 9 V, in which the output increases with temperature. Draw a graph of the output pd with temperature.

5. A new HDTV has a label giving its power rating as 52 W and its annual energy use as 76 kW h. What assumption was made in calculating the annual energy use?

6. Draw a circuit diagram for a light sensor in which the output pd decreases with light level. Explain its operation.

# 2.4 The nature of waves

Water waves, earthquake waves and sound all travel in a similar way, which we shall now study. We shall also begin to study the nature of *light*.

## 2.4.1 How a wave travels

A *progressive wave* (usually just called a *wave*) is a disturbance which travels through a medium. Example: air is a medium for sound waves.

The wave *source* is usually an *oscillating* (vibrating) object in contact with the medium. This source keeps the particles of the medium next to it oscillating. These particles pass the oscillations to their neighbours and so on, so a wave of particle disturbance *propagates* (travels) through the medium, taking *energy* with it.

## 2.4.2 Transverse and longitudinal waves

### (a) Longitudinal waves

Examples: 'compression' waves in a 'Slinky' spring (Fig. 2.4.1), sound, seismic (earthquake) 'P' waves. Learn the definition in *Key terms*: its meaning should be clear from the diagram.

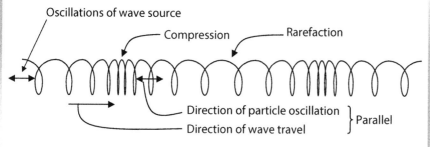

Fig 2.4.1 Longitudinal wave in a 'Slinky' spring

### (b) Transverse waves

Examples are light and other electromagnetic waves, seismic 'S' waves, waves in a taut string (Fig. 2.4.2).

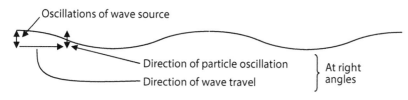

Fig 2.4.2 Transverse wave in a taut string

> **» Pointer**
> When a wave is travelling through a medium, the *particles* of the medium oscillate about their normal positions. They don't really *travel* because they never *get* anywhere!

> **Key Terms**
> In a **longitudinal wave** the particle oscillations are parallel to the direction of travel of the wave.
>
> In a **transverse wave** the particle oscillations are at right angles to the direction of travel of the wave.

> **quickfire**
> ① In the aerial (or antenna) of a radio, free electrons experience oscillating forces from e-m waves. The aerial works best when the forces are parallel to its length. If this is so when the aerial is vertical, are the waves near the aerial *travelling* in a horizontal or vertical direction?

In a **polarised** (strictly **plane polarised**) transverse wave, the oscillations are in just one of the many directions at right angles to the direction of travel.

In an **unpolarised** transverse wave the direction of oscillation keeps changing randomly – but is always at right angles to the direction of travel.

Polarised          Unpolarised

*Fig. 2.4.3 Polarised and unpolarised waves*

The **amplitude**, $A$, of an oscillating particle is the maximum value of its displacement (from its equilibrium position).

The **period**, $T$, of an oscillation is the time taken for one complete cycle.

The **frequency**, $f$, is the number of cycles of oscillation per unit time.

**In phase**, applied to oscillating particles, means at the same point in their cycle at the same time.

**In antiphase** means out of phase by half a cycle, so *always* having displacements in opposite directions.

The **wavelength** is the minimum distance, measured along the direction of travel, between two points in a wave oscillating in phase.

## Electromagnetic waves

Electromagnetic (e-m) waves are special. They don't need a medium, and can travel in a vacuum, where their speed, denoted by $c$, is $3.00 \times 10^8$ m s$^{-1}$. To three significant figures, their speed in air is the same.

For an e-m wave the oscillations are not those of particles, but of *electric* and *magnetic fields*. The wave can't exist without both fields, but we shall not need to refer again to the magnetic field. The oscillating electric field can be detected in this way: a charged particle placed in the path of an e-m wave will experience an oscillating force, in a direction at right angles to the wave's direction of travel.

# 2.4.3 Polarisation

The transverse wave shown in Fig. 2.4.2 is **polarised**, as defined in *Key Terms*. Contrast it with an **unpolarised** wave, which is also defined.

We can sum up the difference using two diagrams (Fig. 2.4.3), drawn for waves travelling out of the page towards you. The arrows are oscillation directions. For an e-m wave they are directions of the oscillating electric field.

The light from most 'normal' sources (including the Sun, a flame, a filament lamp) is unpolarised, but we can polarise it by removing all oscillation components at right angles to one particular direction. This can be done by passing it through a *polarising filter* – Polaroid is often used.

# 2.4.4 Oscillating particles: Displacement–time graphs

Consider a particle of the medium in the path of the wave. Its displacement from its undisturbed position might vary with time as in graph (a) of Fig. 2.4.4.

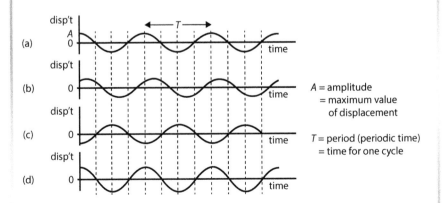

$A$ = amplitude
   = maximum value
    of displacement

$T$ = period (periodic time)
   = time for one cycle

*Fig 2.4.4 Displacement–time graphs*

The **frequency**, $f$, is the number of cycles per unit time. Unit: hertz (Hz). For example, if $T = 0.10$ s, then $f = 10$ Hz. The relationship is

$$f = \frac{1}{T} \quad \text{which can be re-arranged as} \quad T = \frac{1}{f}$$

All the graphs in Fig 2.4.4 are for oscillations of the same frequency.

Graphs (a) and (d) show oscillations that are **in phase**, that is always at the same point in their cycle at the same time. What is the difference between them?

The oscillations in graph (b) are *out of phase* with those in (a). They lag behind by about an eighth of a cycle: their peaks occur further along the time axis.

The oscillations in graph (c) are out of phase with those in (a) by half a cycle. We say that they are in **antiphase**, because *at all times* the displacements are in opposite directions.

## 2.4.5 Snapshot of a wave: Displacement–position graphs

The graph in Fig. 2.4.5 shows the displacement of particles in the medium in the path of a wave *at one particular time*. So the horizontal axis represents position, that is distance from the source.

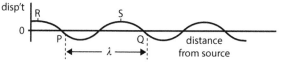

$\lambda$ = wavelength
= distance between consecutive particles oscillating in phase

*Fig 2.4.5 Displacement–distance graph*

Particle Q is oscillating in phase with particle P. Another pair of in phase particles is R and S.

## 2.4.6 Speed of a wave

The snapshot shown by the pecked line in Fig. 2.4.6 shows the wave having advanced to the right by distance $\lambda/4$ from the solid line snapshot.

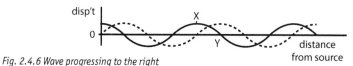

*Fig. 2.4.6 Wave progressing to the right*

This advance must have taken a time $T/4$; for example particle X has gone from peak to zero displacement, and particle Y (further from the source and lagging behind X by a quarter of a cycle) has gone from zero to peak.

The wave will, then, take a time $T$ to move forward by distance $\lambda$.

$$\text{So, wave speed, } v = \frac{\text{distance gone}}{\text{time taken}} = \frac{\lambda}{T} = \frac{1}{T}\lambda = f\lambda.$$

**quickfire**

② Calculate the frequency of the oscillations in Fig. 2.4.5, if the separation between pecked lines represents 20 ms (20 milliseconds).

**quickfire**

③ A particle oscillates at 3000 cycles per minute. Calculate, in SI units, its frequency and its period.

**》 Pointer**

The frequency, $f$, of the wave is always set by the source. The wave speed is usually set by the medium.

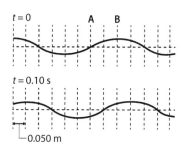

*Fig. 2.4.7 Progressive wave*

### quickfire

④ For the wave in Fig. 2.4.7:
  (a) Calculate:
    (i)   The wavelength
    (ii)  The wave speed
          (assuming it is
          less than
          4.0 m s⁻¹).
    (iii) The frequency.
  (b) For points A and B
      compare:
    (i)   The amplitudes.
    (ii)  The phases.

### Key Term

A **wavefront** is a surface
(or line) on which all
oscillations are in phase.

## 2.4.7 Waves travelling in 2 and 3 dimensions

Many types of wave travel 2-dimensionally (as ripples on a pond appear to do) or 3-dimensionally, like light and sound. We can still use graphs of displacement against distance from source; the amplitude will steadily get less, because the wave's energy is more and more spread out. The spreading is clear in another type of diagram, showing **wavefronts**, as in Fig. 2.4.8

- The curved lines are wavefronts: all particles on any wavefront are oscillating **in phase**. [Wavefronts can sometimes be partly straight.]

- We usually draw wavefronts at intervals of $\lambda$, like the peaks (crests) of a water wave.

- The direction of travel of a wave at any point is at right angles to the wavefront through that point.

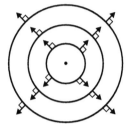

Fig. 2.4.8 Wavefronts from a small source

## 2.4.8 Specified practical work

### (a) Investigating the polarisation of microwaves

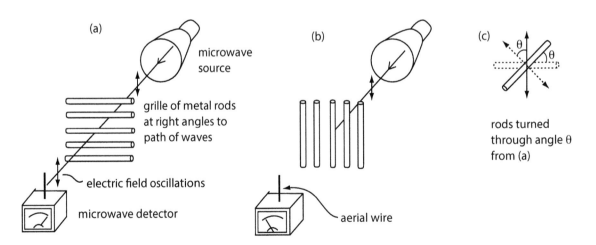

(a)
microwave
source

grille of metal rods
at right angles to
path of waves

electric field oscillations

microwave detector

(b)

(c)
θ
θ
rods turned
through angle θ
from (a)

aerial wire

Fig. 2.4.9 Showing that microwaves are polarised

E-m waves of wavelength a few centimetres (microwaves) can pass freely through a grille of metal rods when the rods are in one of the directions at right angles to the waves' direction of travel. The electric field amplitude of the microwaves getting through gradually falls to zero as the grille is turned through a right angle, from (a) to (b) in Fig. 2.4.9.

In Fig. 2.4.9 (a) the grill lets through the waves to the detector, where their oscillating electric field forces electrons to move up and down the aerial.

In Fig. 2.4.9 (b) the electrons are made to move up and down the rods of the grill, and this results in a reflected wave. Very little wave energy gets through.

This couldn't happen unless the microwaves were transverse, *and* polarised.

We can investigate quantitatively using a signal strength meter. From the position of maximum signal strength, rotate the grill through a succession of angles, $\theta$, measured with a protractor, reading the signal strength, $S$, each time. Plot $S$ against $\cos^2 \theta$, which should show a proportional relationship (see Pointer).

## (b) Investigating the polarisation of light

Polaroid is a man-made material containing long parallel molecules that allow electrons to pass to and fro along them. It serves the same purpose for light ($\lambda \sim 10^{-6}$ m) as the grille of rods does for microwaves ($\lambda \sim 10^{-2}$ m). See Pointer. Study this sequence of experiments.

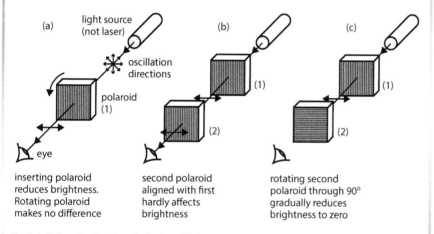

Fig. 2.4.10 Investigating the polarisation of light

The null effect of rotating the single polaroid in Fig. 2.4.10 (a) shows light from an ordinary source to be unpolarised. In (b) and (c) the first polaroid allows through only electric field oscillation components perpendicular to its molecules, so it polarises the light.

We can use a light intensity meter to investigate quantitatively. Starting from the set-up in (b), rotate polaroid (2) through a succession of angles, $\theta$, measured with a protractor, reading the intensity, $I$, each time. Plot $I$ against $\cos^2 \theta$.

## Extra questions

1. Look again at the two graphs in Fig. 2.4.7. Quickfire 4 part (a)(ii) included an assumption.

   (a) Explain why it is necessary to make an assumption in order to answer the question.

   (b) If the assumption that speed $< 4$ m s$^{-1}$ is not made, repeat parts (a)(ii) and (iii) for the next two possible values of speed.

2. A transverse wave of frequency 50 Hz is travelling to the right as shown in the snapshot (right).

   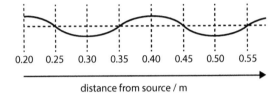

   (a) Calculate the speed of the waves.

   (b) How would a snapshot taken 0.005 s later compare with the one on the right?

   (c) Explain why the waves are called *transverse*.

3. A strip of wood, in contact with the surface of water in a tank, oscillates up and down at a frequency of 5.0 Hz. The view from above shows the positions of wave crests at one instant. Note the distance scale.

   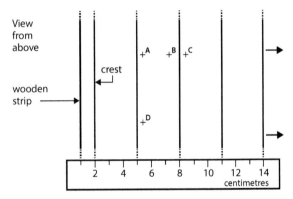

   (a) Calculate the time it would take a wave crest to travel a distance of 10.5 cm.

   (b) State, giving a reason, whether or not the oscillations at B, C and D are *in phase* with the oscillations at A.

4. The two graphs, **A** and **B**, are of the same wave. Graph **A** is a snapshot of the wave at a particular time. Graph **B** is a displacement–time graph at a particular position.

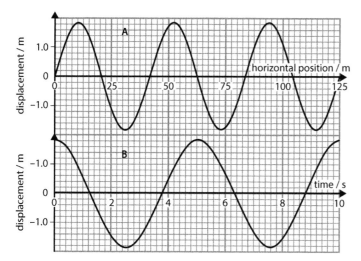

(a) Use the graphs to determine:

(i) the wavelength (ii) the frequency and (iii) the speed of the wave.

(b) If graph **A** is the snapshot at time $t = 5.0$ s, give three possible positions of graph **B.**

(c) Calculate the maximum vertical speed of a point on the wave.

# 2.5 Wave properties

We deal here with diffraction and – perhaps the most interesting thing that waves do – interference. Diffraction is treated as a phenomenon (something that happens), but you are required to understand *why* interference happens in the way that it does.

## 2.5.1 Diffraction

When waves arrive at a barrier with a slit in it, some pass through the gap and, to some extent, spread out. This is a case of **diffraction**.

### (a) Slit width ≤ λ

In this case the waves spread right round though 90° each side of the straight-through direction. See wavefront diagram (Fig. 2.5.1). Note that the amplitude falls off at the sides.

The waves spread out rather as if they came from a small wave source in the slit itself.

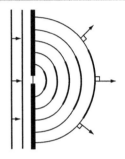

Fig. 2.5.1 Diffraction: Slit width ≤ λ

### (b) Slit width > λ

In this case there is a main or central beam that doesn't spread all the way round. There are also 'side beams' of much lower amplitude.

The wider the slit, the greater the amplitude of the main beam, but the less its angular spread.

In fact there is very little diffraction when the wavelength is much smaller than the width of the obstacle or slit. Here is an example.

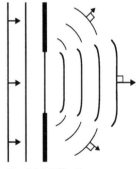

Fig. 2.5.2 Diffraction: Slit width > λ

### (c) Demonstrating the diffraction of light

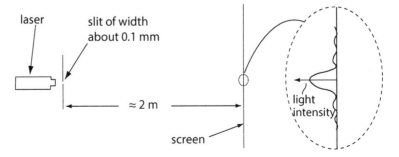

Fig. 2.5.4 Demonstrating the diffraction of light through a slit

**» Pointer**

It is also diffraction when waves in the sea come together on the far side of a rock. (Fig. 2.5.3)

Fig. 2.5.3 Diffraction round a rock

 **Grade boost**

When drawing wavefront diagrams, don't let λ change (unless the medium changes – see Section 2.6)

**quickpire**

① Why can you hear someone through an open door, when you can't see them? [Hint: a typical wavelength for sound is 1 m.]

Where the intensity is greatest, so is the wave amplitude. Note the exaggerated scale of distances up and down the screen; the light spreads only through a very small angle. This is because the maximum wavelength of light (extreme red) is about 700 nm, so the slit is over 100 wavelengths wide.

## quickᖴire

② Suggest why reception of FM radio signals (frequency around 100 MHz) is often poor in deep narrow valleys, whereas for 'long wave' AM (frequency <1 MHz) reception is fine. [Radio waves are e-m waves.]

## 2.5.2 Interference pattern of waves from two sources

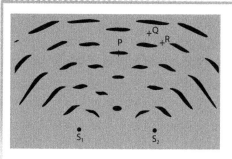

Fig. 2.5.5 shows wavefronts arising from $S_1$ and $S_2$, two in-phase sources (sources oscillating in phase). Note the 'beams' of high amplitude separated by 'channels' of zero (or almost zero) amplitude.

*Fig. 2.5.5 Two source interference pattern*

The pattern (based on a photograph of water waves from a two-pronged oscillating dipper) can be explained using **the principle of superposition**.

>> *Pointer*

*'Vector sum'* implies that equal and opposite displacements will add to zero.

This can't happen for two displacements at right angles. So, for polarised waves, the vibration cannot be at right angles if the waves are to produce an interference pattern.

## 2.5.3 Constructive and destructive interference

Where the amplitude is highest (for example at P or Q in Fig. 2.5.5) the waves from both slits are arriving in phase and interfering **constructively**, as shown in Fig. 2.5.6(a), in which the principle of superposition is applied.

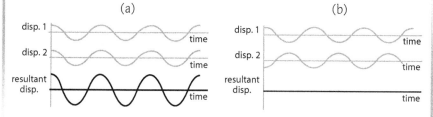

*Fig. 2.5.6 Demonstrating the diffraction of light through a slit*

Where the amplitude is lowest (for example at R) the waves arrive in antiphase, and interfere **destructively** as shown in Fig. 2.5.6 (b).

>> *Pointer*

Q in Fig. 2.5.5 *does* have maximum amplitude, even though the displacement may not be at its greatest at the instant for which the diagram is drawn.

**quickfire**

③ For R in Fig 2.5.5 what is the path difference $S_1R$ in terms of wavelength? [Hint: R is to the right of Q.]

>> **Pointer**

The path difference rules in 2.5.4 need to be known (and understood!). They are not in the Data Booklet.

# 2.5.4 Path difference

It's easy to understand *why* there must be constructive interference at P in Fig. 2.5.5: in order to get to P, the waves from $S_1$ and $S_2$ travel along **paths,** $S_1P$ and $S_2P$, of *equal length*, and therefore arrive at P in phase.

For point Q, the paths are $S_1Q$ and $S_2Q$. The **path difference,** $S_1Q - S_2Q$, is exactly 1 wavelength, *so* waves from $S_1$ arrive at Q one whole cycle later than those from $S_2$, which means they arrive at Q in phase with those from $S_2$.

The general rule, for waves from in-phase sources, is:

For constructive interference at a point X,

Path difference, $S_1X - S_2X = 0$, or $\lambda$, or $2\lambda$, or $3\lambda$ ...

For destructive interference at a point X,

Path difference, $S_1X - S_2X = \frac{1}{2}\lambda$, or $\frac{3}{2}\lambda$, or $\frac{5}{2}\lambda$ ...

>> **Pointer**

Making measurements from the set-up in Fig. 2.5.7 is discussed in Section 2.5.9.

# 2.5.5 Young's *Fringes* experiment

In the early 1800s, Thomas Young investigated the light passing through two parallel slits close together. He observed a pattern of bright and dark *fringes* (stripes) on a screen placed some distance in front of the slits. He recognised this as part of an interference pattern, deduced that light was wave-like, and determined the wavelengths of different colours of light.

Fig. 2.5.7 shows a modern version of Young's experiment. Typically, $a = 0.50$ mm, $D = 2.0$ m, and we find that for red light, $y$ is roughly 2.5 mm. Note the exaggerated vertical scale on the graph.

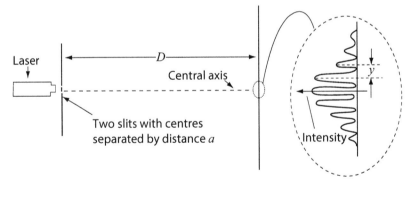

**quickfire**

④ Why does the intensity of the bright fringes fall off with distance from the central axis? [Hint: at yet greater distances, fringes re-appear, but faintly.]

*Fig. 2.5.7 Young's fringes experiment*

## (a) The role of diffraction in Young's experiment

Light diffracts slightly on passing through each slit, so there is a (smallish) region where the light from each slit overlaps – and interferes.

## (b) The equation for Young's experiment

The wavelength of the light can be found from the equation

$$\lambda = \frac{ay}{D}$$

This is an approximation based on the path difference rule for constructive interference. As long as $a \ll D$ and $y \ll D$ the equation is almost exact.

## (c) Accurate measurement of wavelength

Physicists need to measure wavelengths accurately, for example when trying to identify atoms in a star's atmosphere by the light they absorb and emit. The Young's fringes set-up would be too inaccurate for three reasons:

- The fringes are not sharp: bright fringes fade gradually into dark.
- The 'bright' fringes are therefore not very bright.
- The fringe separation is small.

All three issues are addressed by the **diffraction grating** (Section 2.5.7).

# 2.5.6 Conditions for observing interference; coherence

A laser is an ideal source for illuminating the slits in Young's experiment, because it produces **coherent** light (see **Key terms**.). This ensures that the slits act as in-phase sources or, if the laser beam is not pointing exactly head-on at them, there is at least a *constant phase relationship* between them. We could say that the sources are 'mutually coherent'.

An 'ordinary' light source such as an LED or a filament lamp is an **incoherent** (non-coherent) **source**. It won't give fringes if placed where the laser was placed (Fig. 2.5.7); there won't be a constant phase relationship between the light from the two slits. [It is *possible* to obtain fringes using light derived from an ordinary source if a special arrangement is used for illuminating the slits. It doesn't have to be very special – just a very narrow light source, e.g. one with a planar filament viewed from the side.]

Not surprisingly, if we illuminate one slit in Young's experiment with light from one ordinary source (for example an LED), and the other slit with light from a different source (even an identical LED), we can *never* produce fringes.

As explained in Section 2.4.3 (Pointer), if the light is polarised, the directions of electric field oscillations for light from the two slits cannot be at right angles, for destructive and constructive interference to take place.

## quickfire

⑤ Why *must* two of the quantities $a$, $y$, $D$, be on the top line on the right-hand side of the equation, and one on the bottom, rather than vice versa?

## quickfire

⑥ Suppose $a = 0.50$ mm, $D = 2.0$ m, and $y$ is found to be 2.5 mm. Calculate the wavelength of the light.

## Key Terms

**Coherent light** is monochromatic and has wavefronts stretching across the width of its beam, as though it came from a point source.

Two (or more) sources are (mutually) **coherent** when there is a constant phase relationship between them.

# 2.5.7 The diffraction grating

## (a) What is a diffraction grating and where does diffraction occur?

At its simplest, a diffraction grating is a flat plate which is opaque except for thousands of narrow, straight, parallel, equally-spaced slits.

We shine light normally (at right angles) on to the grating, so each wavefront reaches every slit at the same time. The slits therefore act as in-phase sources and each, being only a few wavelengths in width, sends out waves that spread widely.

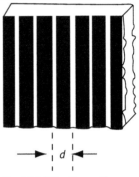

Fig. 2.5.8 Diffraction grating

### quicKpire

⑦ Calculate $d$ for a grating with 500 slits per millimetre of its width.

## (b) Beams and orders

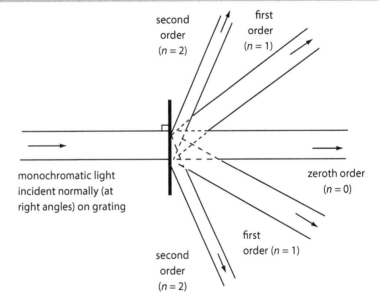

Fig. 2.5.9 Beams and orders

>> **Pointer**

There may be more or fewer orders than those shown in Fig. 2.5.9. See **Example**.

**Grade boost**

Note the way the beams are labelled in Fig. 2.5.9. If the highest order beam is the second ($n = 2$), then 5 beams will emerge. Don't confuse number of orders with number of beams.

**Key Terms**

**Monochromatic light** is light of a single frequency (in practice a narrow range of frequencies).

The diffracted light from different slits interferes to produce beams. They correspond to those giving the bright fringes in Young's experiment, but are much further apart, because the slits are much closer together. The beams are also bright and well-defined, separated by dark areas of nearly total destructive interference.

When the beams meet a screen they produce bright spots in a line. If the light source is a distant slit, parallel to the grating slits, illuminated with **monochromatic light**, the dots are now lines, so we say that a monochromatic light source has a 'line spectrum'.

## (c) The grating equation

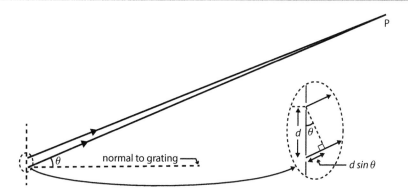

Fig. 2.5.10 Path difference for a diffraction grating

**》 Pointer**

If the condition
$n\lambda = d \sin\theta$ [n = 0 or 1, 2,....] is met, light from *each* slit interferes constructively with light from its neighbour, so light from *all* slits will interfere constructively, hence the brightness of the beams.

When the condition is *not* met, there are more possibilities for destructive interference than with only two slits, hence the sharpness of the bright beams.

The paths of light from adjacent slits to a distant point, P, will be almost parallel. Dropping a perpendicular as in Fig. 2.5.10 gives a right-angled triangle that includes angle $\theta$, equal to the angle between the light paths and the normal to the grating. We see that for light going to P from adjacent slits,

Path difference = $d \sin\theta$

This must equal zero, or a whole number of wavelengths, for constructive interference at P. So for bright beams,

$n\lambda = d \sin\theta$    [n = 0 or 1, 2, 3....]

For any value of $\lambda$, n = 0 gives $\theta = 0$, so n = 0 corresponds to the zeroth order. n = 1 corresponds to the first order and so on.

**》 Pointer**

$\theta$ is the angle between beam and normal to grating.

### Example

A diffraction grating has 400 slits per millimetre. Light from a street lamp falling normally on it produces third-order beams at 45° either side of the normal. Determine:

(a) The wavelength of the light.

(b) The highest order produced.

### Answer

(a) First we note that since there are 400 lines per millimetre, then

$$d = \frac{1}{400} \text{ mm} = 2.5 \times 10^{-3} \text{ mm} = 2.5 \times 10^{-6} \text{ m}$$

Now we can use the grating equation:

$$\lambda = \frac{d \sin\theta}{n} = \frac{2.5 \times 10^{-6} \text{ m} \sin 45°}{3} = 5.9 \times 10^{-7} \text{ m} = 590 \text{ nm}$$

(b) One way to do this is to re-arrange the grating equation as $n = \frac{d \sin\theta}{\lambda}$.

Now $\sin\theta$ can't be greater than 1, so $n \le \frac{d}{\lambda}$.

In this case, $n \le \frac{2.5 \times 10^{-6} \text{ m}}{5.9 \times 10^{-7} \text{ m}}$. Doing the division, we find $n \le 4.2$, so the maximum order number is 4.

**quickfire**

⑧ Monochromatic light is shone normally on to a grating with 500 slits per mm. The second order beams are at 36° to the normal. Calculate:

(a) The wavelength.

(b) The highest order number for this wavelength.

**quickfire**

⑨ A laser beam is shone normally at a grating with $5.00 \times 10^5$ slits per metre; 7 beams emerge. The angle between the outer beams is 144°. Calculate the wavelength.

*Hints*: Use the equation but beware: n ≠ 7 and $\theta$ ≠ 144°.

# 2.5.8 Stationary (standing) waves

These can be set up on a stretched string as shown in Fig. 2.5.11, and viewed in slow motion with a stroboscope (a light flashing at regular intervals).

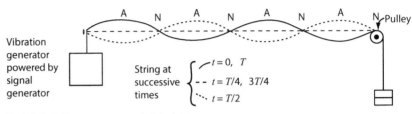

Fig. 2.5.11 Stationary waves on a stretched string

Three snapshots, at intervals of $T/4$ (a quarter of a cycle), are superimposed in Fig. 2.5.11. They should help you understand the description (definition) of a stationary wave given in **Key terms**.

Contrast a stationary wave with a progressive wave (Section 2.4), for which:

- the amplitude doesn't change, or may fall off steadily, with distance from the source. there are no nodes or antinodes,
- the phase lags at a constant rate with distance, one cycle per wavelength,
- energy is transported with the wave.

## (a) Stationary waves as interference patterns

When two progressive waves of equal amplitude and frequency, travel in opposite directions in the same region of space, they interfere to produce a stationary wave.

In the set-up shown in Fig. 2.5.11 a progressive wave from the vibration generator is travelling to the right, and a (reflected) wave from the pulley, to the left.

Antinodes correspond to constructive interference. They are spaced $\lambda/2$ apart. See Pointer. Nodes (destructive interference) are also $\lambda/2$ apart.

## (b) Stationary waves on taut strings fixed at both ends

Since there must be nodes at each end, the string can vibrate in only certain *modes*. The three lowest modes are shown in Fig. 2.5.12.

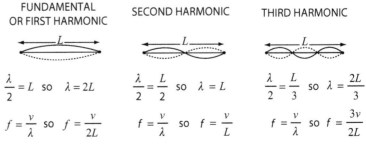

Fig. 2.5.12 Stationary wave modes on a string

For Fig. 2.5.12 modes:

FUNDAMENTAL OR FIRST HARMONIC

$$\frac{\lambda}{2} = L \text{ so } \lambda = 2L$$

$$f = \frac{v}{\lambda} \text{ so } f = \frac{v}{2L}$$

SECOND HARMONIC

$$\frac{\lambda}{2} = \frac{L}{2} \text{ so } \lambda = L$$

$$f = \frac{v}{\lambda} \text{ so } f = \frac{v}{L}$$

THIRD HARMONIC

$$\frac{\lambda}{2} = \frac{L}{3} \text{ so } \lambda = \frac{2L}{3}$$

$$f = \frac{v}{\lambda} \text{ so } f = \frac{3v}{2L}$$

$v$ is the speed of transverse waves along the string. [This depends on the mass per unit length of string, and the tension it is under.]

In *which* mode will the string vibrate? This is determined by whether the string is hit, plucked or bowed, and *where* on the string this is done. In fact the string will usually vibrate in a *sum* of modes.

### (c) Other types of stationary waves

Any type of progressive wave can have a corresponding stationary wave.

For stationary sound waves in pipes, there are nodes of displacement at closed ends and antinodes of displacement at open ends. These stationary waves are responsible for the sounds produced by wind instruments.

## 2.5.9 Specified practical work

### (a) Determination of wavelength using Young's double slit experiment

You need to have learned the basic set-up, as described in Section 2.5.5. We are now concerned with practical details.

#### Safety precautions

We shall assume that the light source is to be a low power ($\leq$ 5 mW) laser 'pointer', though a different source might be provided. If using a laser...

- Do not look into the beam directly or by reflection from a shiny surface.
- Try to ensure that others cannot do so. Never point a laser at anyone.
- Work with usual room lighting (so your pupils are not enlarged).

#### The slits themselves

Two parallel slits, each of width ~0.1 mm, are needed in an otherwise opaque surface. The separation, $a$, of the centres of the slits should be $\leq$ 1 mm. It is possible to *make* the slits, and to measure $a$ using a *travelling microscope*. We shall assume that the slits have been bought from a supplier, with the value of $a$ given (ideally with its uncertainty).

#### Measuring fringe separation

The apparatus should be set up as in Fig. 2.5.7. The distance, $D$, from slits to screen needs to be chosen to give the best chance of accurately measuring the distance, $y$, between the centres of the fringes – see Quickfire 14(a). $D$ itself should be measured with a tape measure or metre rule(s).

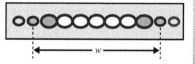

*Fig. 2.5.13 Young's fringes*

**quickfire**

⑩ The fundamental frequency of a string of length 1.50 m is 128 Hz. Calculate the speed of transverse waves on the string.

**quickfire**

⑪ If a string fixed at both ends is set vibrating by hitting it in the middle, which harmonics won't be present?

**quickfire**

⑫ Give too reasons why $a$ must not be too large.

**quickfire**

⑬ (a) Why should $D$ not be too small?
Recall that
$$y = \frac{D\lambda}{a}.$$
(b) Apart from room size, what puts an upper limit on $D$?

**quickfire**

⑭ (a) Suppose that $w$ (see Fig. 2.5.13) is measured to be 15.0 mm ± 0.5 mm. Determine $y$, together with its percentage uncertainty.
(b) $D$ is measured to be 1500 ±5 mm and $a$ is given as 0.50 mm ± 0.02 mm. Calculate the wavelength of the light and its absolute uncertainty.

Measure the distance, $w$, between the centre of the $n$th fringe and that of the $n$th fringe away; hence $y = w/2n$.

### Calculating the wavelength

We use the equation $\lambda = \dfrac{ay}{D}$. See Quickfire 14(b).

### Extending the investigation

You could investigate the dependence of $y$ on $D$, plotting the appropriate graph. Or you could observe and interpret white light fringes; a suitable light source is a straight lamp filament placed a metre or two from the slits and aligned parallel to them.

## (b) Determining wavelength using a diffraction grating

When used with an instrument called a *spectrometer*, a diffraction grating delivers very accurate values for wavelength. Fig. 2.5.14 shows a set-up that doesn't use the grating's full capabilities, but is very simple.

First make sure you have learnt the basics (Section 2.5.7).

### Safety precautions

See previous experiment (Young's fringes) if using a laser 'pointer'.

### Determining beam angles and hence wavelength

$D$ should be chosen to minimise uncertainties in the angles – determined by measuring distances $D$, $B_1$, $B_2$, (and $B_3$ if third order present), using a metre rule.

Then
$$\theta_1 = \tan^{-1}\frac{B_1}{2D'}$$

$$\theta_2 = \tan^{-1}\frac{B_2}{2D'}$$

$$\left[\theta_3 = \tan^{-1}\frac{B_3}{2D}\right]$$

Substituting these values into the grating equation (for $n = 1, 2, [3]$) should give the same value for $\lambda$ each time. Hence a 'best' value.

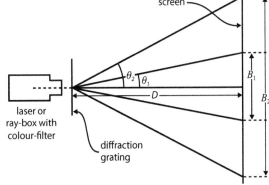

Fig. 2.4.14 Simple diffraction grating set-up

> ## » Pointer
> The angles could be measured directly with a protractor by placing a screen in the plane of the apparatus as drawn, so that the beams brush against it. You should consider whether this would be more or less accurate than the method suggested.

### quickfire

⑮ A grating with slit separation 2000 nm is used in the set-up shown, with $D = 400$ mm. $B_1$ and $B_2$ are measured to be 220 mm and 505 mm respectively.

(a) Calculate a value for $\lambda$.

(b) Suggest why $B_3$ is hard to measure. How else might $\theta_3$ be measured?

## (c) Determining speed of sound in air using stationary waves

Here we set up stationary sound waves (Section 2.5.8) in an air 'column' in a tube. In the version shown in Fig. 2.5.15 the 'floor height' can be adjusted gradually by releasing water from the tube.

### The fundamental mode in the air column

This is when there is a node (N) at the 'closed end' (the water surface), an antinode (A) just beyond the open end (distance $b$ above top of tube), and no nodes or antinodes in between.

$$\text{So } \frac{\lambda}{4} = L_1 + b \quad \text{so} \quad L_1 = \frac{v}{4f} - b$$

Here we have used $v = f\lambda$ in which $v$ is the speed of sound in air, and $f$ is the fundamental frequency of the air column.

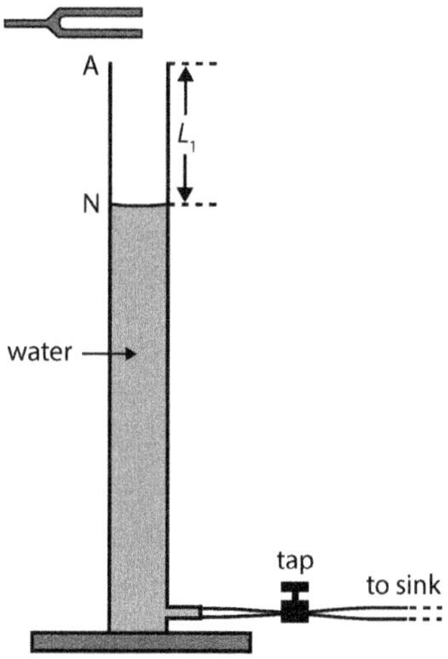

Fig. 2.5.15 Resonance tube

### Finding resonance in the fundamental mode

If a tuning fork is struck gently and placed a little way above the top of the tube, a louder sound will be heard for a particular value of $L_1$ as $L_1$ is slowly increased. See Pointers. This effect is called *resonance*. It occurs when the frequency of the stationary wave equals that of the fork – as marked on the fork.

### Suggested treatment of results

Values of $L_1$ for resonance can be found for five or six different frequencies, $f$, using different tuning forks (or signal generator settings). Plot $L_1$ against $\frac{1}{f}$.

Comparing $L_1 = \frac{v}{4f} - b$ with $y = mx + c$ we see that the gradient is $\frac{v}{4}$ and the intercept is $b$.

An alternative procedure uses a single tuning fork, but we determine not just $L_1$, but also $L_3$, the length of air column that gives the same frequency of stationary wave when there is an antinode (A) and a node (N) *between* the closed and open ends, giving the sequence NANA instead of NA.

In this case $\frac{3\lambda}{4} = L_3 + b$ so $L_3 = \frac{3v}{4f} - b$. See Quickfire 16.

> ### Pointer
> You'll need to 'refresh' the tuning fork by striking it again from time to time. Instead of the fork, a small loudspeaker connected to a signal generator could be used; adjust it so that the sound is just audible.

> ### Pointer
> It's a good idea to predict a rough value for $L_1$ before trying to find resonance. For example, with a 512 Hz tuning fork, knowing that the speed of sound in air is about 340 m s$^{-1}$,
> $$\frac{\lambda}{4} = \frac{340}{4 \times 512} \text{ m} = 0.17 \text{ m},$$
> so we'd listen for resonance when $L_1$ is between 0.15 m and 0.17 m, as $b$ is roughly half the tube radius: pretty small.

### quickfire

⑯ Using a tuning fork of frequency 512 Hz, $L_1$ and $L_3$ are found to be 0.157 m ± 0.003 m and 0.489 m ± 0.003 m. Calculate a value for the speed of sound, and its uncertainty. [Hint: subtract the equation for $L_1$ from that for $L_3$, to eliminate $b$.]

# Extra questions

1. In a version of Young's double slit experiment, a laser is shone at slits with centres separated by 0.50 mm. The screen is 1.5 m away from the slits. The separation between the centres of adjacent fringes is 2.00 mm.

   (a) Calculate a value for the wavelength of light from the laser.
   (b) Light diffracts at each slit.
   - (i) What does this statement mean?
   - (ii) Explain why diffraction is essential for interference fringes to occur.
   (c) Explain in terms of path difference and phase how dark fringes are produced. [Assume the slits to act as in-phase sources.]
   (d) State *two* ways in which the appearance of the fringes on the screen would change if the distance from the double slits to the screen were increased to 7.5 m.

2. (a) The label indicating slit separation (the separation between the centres of slits) on a diffraction grating has been removed. A student decides to determine it by shining a laser normally at the grating. The wavelength of the laser light is known to be 532 nm.

   He measures the angle between a second order emerging beam and the central (zeroth order) beam to be 28.9°. Show that the slit separation is approximately $2 \times 10^{-6}$ m.

   (b) The student now uses the grating to determine the wavelength of the light from another laser. He measures the second order beam to be at an angle of 35.1°.

   - (i) Calculate the wavelength of the light.
   - (ii) Determine the number of bright beams that the grating produces with this wavelength.

3. The diagram shows a *stationary* wave on a stretched string at a time of maximum displacement.

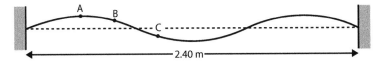

   (a) Determine the wavelength.
   (b) The frequency is 50 Hz. Determine the minimum time taken for the string to change from its position of maximum displacement to the position shown as a pecked straight line.
   (c) Compare (i) the amplitude, (ii) the relative phase of points A and B on the string.
   (d) Compare (i) the amplitude, (ii) the relative phase of points B and C (equal distances either side of a node) on the string.
   (e) Calculate the frequency of the lowest frequency (fundamental) stationary wave on the same string.

4. A student uses an unlabelled diffraction grating, **G**, and a laser, **L**, of wavelength 659 nm to measure the wavelength of a second laser, **M**. Her set-up is shown.

   She measures the separation, $d$, between the two $n$th order spectra with **L** and **M** and the distance $D$ from the grating to the wall.

   (a) Results: $D = 1500$ mm;     $d_L = 365$ mm; $d_M = 270$ mm.
       Calculate the wavelength of **M**.

   (b) A second student forgot to write down the value of $D$ but remembered that $D \gg d$ so that the approximation $\sin \theta \approx \tan \theta$ is quite good. What was this student's value of $\lambda_M$?

# 2.6 Refraction of light

In this section we are concerned with waves passing from one medium to another. We shall be particularly concerned with the behaviour of light.

## 2.6.1 Change of medium: speed, frequency and wavelength

The *speed* of a wave is usually determined by the medium in which it is travelling. For example, at 20°C, sound travels 4.3 times faster in water than in air.

Suppose a wave passes from medium 1, in which its speed is $v_1$ to medium 2 (speed $v_2$).

Its frequency will stay the same (that of the wave source), as each cycle is passed across the interface between 1 and 2. Using $v = f\lambda$

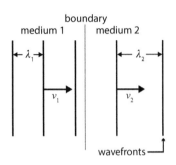

*Fig. 2.6.1 Speed change*

$$f = \frac{v_1}{\lambda_1} = \frac{v_2}{\lambda_2}.$$

## 2.6.2 Refraction of waves

If a wave passes obliquely from one medium to another, in which it travels at a different speed, its direction of travel changes. This is called **refraction**.

In Fig. 2.6.2 (a) waves go from medium 1 to medium 2 ($v_2 > v_1$) and the direction of travel bends away from the normal (a line at right angles to the boundary).

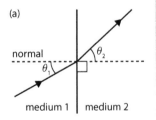

Fig. 2.6.2 (b) shows a wide 'beam' of waves. Wavefronts AB and CD are drawn at right angles to the directions of travel in medium 1 and medium 2. In fact CD is a later position of AB.

We can now see why the direction of travel *must* change in this way. While end A of AB is still travelling at speed $v_1$ to C, end B is travelling at speed $v_2$ to D, so BD > AC.

Taking this further, suppose $t$ is the time for AB to reach CD. Then AC = $v_1 t$ and BD = $v_2 t$. Using the two-right angled triangles, ABC and DCB,

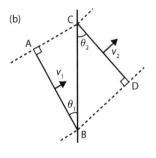

*Fig. 2.6.2 Refraction*

$$\sin \theta_1 = \frac{AC}{BC} = \frac{v_1 t}{BC} \quad \text{and} \quad \sin \theta_2 = \frac{BD}{BC} = \frac{v_2 t}{BC}$$

Then, by division: $\dfrac{\sin \theta_1}{\sin \theta_2} = \dfrac{v_1}{v_2}$

>> **Pointer**

It's not hard to show that the angles marked $\theta_1$ in Figs 2.6.2 (a) and (b) are equal. The same goes for the angles $\theta_2$.

## 2.6.3 Refraction of light

Light travels more slowly in (transparent) solids and liquids than in a vacuum. We define the **refractive index**, $n$, of a medium as

$$n = \frac{\text{speed of light in a vacuum}}{\text{speed of light in the medium}}, \quad \text{that is } n = \frac{c}{v}$$

*Examples*: for air, $n = 1.00$; for water, $n = 1.33$; for ordinary glass, $n \sim 1.5 - 1.7$.

Because $nv = c$, if we compare two media (medium 1 and medium 2):

$$n_1 v_1 = n_2 v_2, \quad \text{that is } \frac{v_1}{v_2} = \frac{n_2}{n_1}$$

We can now write the refraction equation from the previous section in terms of $n_1$ and $n_2$:

$$\frac{\sin \theta_1}{\sin \theta_2} = \frac{v_1}{v_2} \text{ is equivalent to } \frac{\sin \theta_1}{\sin \theta_2} = \frac{n_2}{n_1}$$

Or, memorably: $n_1 \sin \theta_1 = n_2 \sin \theta_2$

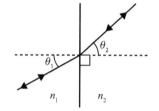

Light paths are reversible (see Fig. 2.6.3, which is drawn for $n_1 > n_2$) so the equation works whether light is going from medium 1 to medium 2 or vice versa.

*Fig. 2.6.3 Reversibility*

### Example

A laser beam is shone into a fish-tank through one of its glass walls, as in Fig. 2.6.4. Angle $\theta_a$ is 45°. Determine the angles $\theta_g$ and $\theta_w$.

[Note that the two angles marked $\theta_g$ are equal only because the glass wall of the tank has parallel faces.]

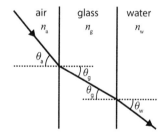

### Answer

At the air–glass boundary:

*Fig. 2.6.4 Refraction into a fish-tank*

$$n_a \sin \theta_a = n_g \sin \theta_g \quad \text{so} \quad 1.00 \sin 45° = 1.52 \sin \theta_g$$

So $\sin \theta_g = \dfrac{1.00 \sin 45°}{1.52}$ so $\theta_g = \sin^{-1} 0.465 = 28°$

At the glass–water boundary,

$$n_g \sin \theta_g = n_w \sin \theta_w \quad \text{so} \quad 1.52 \sin \theta_g = 1.33 \sin \theta_w$$

So $\sin \theta_w = \dfrac{1.00 \times 0.465}{1.52} = 0.532$ so $\theta_w = \sin^{-1} 0.532 = 32°$

### Key Terms

The **refractive index**, $n$ of a transparent material is defined by $n = \dfrac{c}{v}$, in which $c$ is the speed of light in a vacuum and $v$ is its speed in the material.

**Snell's law**

For a light beam passing from one medium to another

$$n_1 \sin \theta_1 = n_2 \sin \theta_2$$

in which $\theta_1$ and $\theta_2$ are angles of beams to the normal. $n_1$ and $n_2$ are constants, the refractive indices, for the materials.

>> **Pointer**

A nice way to remember Snell's law which is **not** in the Data Booklet is:
$$n \sin \theta = \text{constant}$$

⚠ **Grade boost**

Your calculator needs to be in *degrees* rather than *radians* mode. To check that you are finding inverse sines correctly, try $\sin^{-1} 0.5$. It should be 30°.

>> **Pointer**

The path of light approaching a surface is called the incident ray. The angle between *the incident ray* and the normal is *the angle of incidence* (e.g. $\theta_a$ in Fig. 2.6.4).

## quickfire

③ Light travelling through air is incident on a water surface at 45°. Calculate the angle of refraction, and compare it with $\theta_w$ in the example. Also see Extra question 6.

### Sketching paths of refracted light

Draw a normal (dashed or pecked) wherever the light beam hits a boundary as in Fig. 2.6.5. Light bends away from the normal when it goes into a medium of smaller refractive index (where it travels faster), and towards the normal when it goes into a medium of greater refractive index. The light-paths are often called 'rays'.

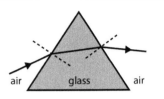

*Fig. 2.6.5 Ray sketching*

# 2.6.4 Critical angle and total internal reflection (TIR)

Consider a light beam travelling in a medium (1) and approaching a medium (2) with a smaller refractive index, in which the light travels faster (e.g. light approaching air from glass).

In Fig. 2.6.6 (a), where $\theta_1$ is smallish, the light bends away from the normal as expected.

Note that some of the light is reflected back into medium 1 (partial reflection). The angles of incidence and reflection (that is the angles of the beams to the normal) are equal.

If we gradually increase $\theta_1$, it is going to reach a value, $\theta_C$, called **the critical angle** at which $\theta_2$ is 90°. See Fig. 2.6.6 (b).

In this case    $n_1 \sin \theta_C = n_2 \sin 90°$

But $\sin 90° = 1$   so   $n_1 \sin \theta_C = n_2$

When $\theta_1 > \theta_C$ the refraction equation gives $\sin \theta_1 > 1$, which is absurd as there is no angle with a sine greater than 1. The equation simply *does not apply* when $\theta_1 > \theta_C$. In fact *no* light enters medium 2. Instead the light is **totally internally reflected** (at an equal angle) back into medium 1 – Fig. 2.6.6 (c).

Total internal reflection is the effect that gives diamonds their sparkle. It is also used instead of mirror reflection in high-class optical instruments such as prism binoculars. In the next section we home in on its use in fibre optics.

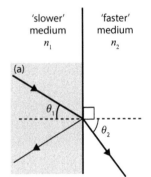

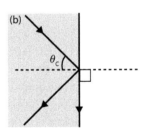

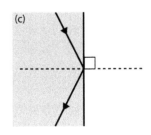

*Fig. 2.6.6 Refraction and reflection*

## Key Terms

**The critical angle**, $\theta_C$, is the angle of incidence for light approaching a 'faster' medium from a 'slower' one, at which the angle of refraction is 90°.

**Total internal reflection** is the total reflection of light that approaches a '*faster*' medium from a '*slower*' medium at an angle of incidence greater than the critical angle.

## ≫ Pointer

The reflection is said to be *internal* because medium 1 is usually a solid or liquid and medium 2 is often just air.

## Example

Light approaches the clear plastic prism as shown in Fig. 2.6.7 (a). Show its path through the prism and out into the air, making the necessary calculation. The refractive index of the plastic is 1.49.

## Answer

The light enters the prism along the normal, so no bending at the first surface, BC. It strikes CA an angle of incidence of 45°. Can it get out? We calculate the critical angle...

$$1.49 \sin \theta_c = 1.00 \sin 90°$$

so $$\sin \theta_c = \frac{1}{1.49}$$

and $$\theta_c = \sin^{-1} \frac{1}{1.49} = 42°$$

Since 45° > 42° there is total internal reflection at CA. The same applies at AB, so the path is as shown in Fig. 2.6.7 (b).

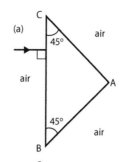

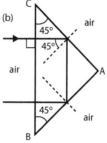

Fig. 2.6.7 Prism example

**quickfire**

④ Calculate the critical angle for light approaching water ($n = 1.33$) from glass ($n = 1.52$).

**quickfire**

⑤ Repeat the example for a prism made out of ice ($n = 1.31$). Show that the light ray does not suffer TIR and sketch its path through the prism and out into the air.

See also Extra question 7.

# 2.6.5 Fibre optics

An optical fibre is a long, thin cylindrical **core** of glass, encased in a **cladding** of glass of lower refractive index.

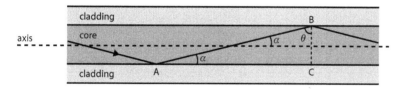

Fig. 2.6.8 Total internal reflection in an optical fibre

Suppose light enters the core at one end, and travels at an angle $\alpha$ to the axis. If $\alpha$ is small enough, the light will strike the core/cladding boundary at an angle $\theta$ to the normal ($\theta = 90° - \alpha$) greater than the critical angle. Then the light will totally internally reflect back into the core and will carry on along the fibre in a zigzag path without escaping. This happens even if the fibre is bent into a (gentle) curve. Optical fibres can therefore take light to hard-to-reach places (like the stomach).

## Carrying data in optical fibres

Data can be *encoded*, as a sequence of pulses, in light (often from a laser) and sent along a fibre. In very pure glass the light 'signal' travels many kilometres without needing a boost – unlike a signal sent by electric currents in copper wires. These also suffer 'cross-talk' from other wires, whereas fibres (in opaque sheaths) are almost immune to outside influences.

## quickfire

⑥ A multimode fibre 250 m long transmits a light pulse in a time of 1.28 μs by the shortest path.

  (a) Show that the refractive index of the core is 1.54 (3 s.f.).

  (b) Calculate the angle to the axis of the longest totally internally reflecting path if the light takes 1.31 μs by this path.

  (c) Explain why light cannot travel far at a greater angle to the axis than the one you have calculated.

## quickfire

⑦ (a) Show that the speed of light in the optical fibre core is $2.00 \times 10^8$ m s$^{-1}$.

  (b) Calculate the time for the shortest path through the optical fibre.

# 2.6.6 Multimode dispersion

One type of fibre used in data transmission is called a **multimode fibre**, because each pulse can travel by several different paths at once. The shortest path is straight along the fibre axis; the other paths are zigzags. The longest is at an angle $\alpha$ to the axis such that $\theta = \theta_c$. Because the paths are of different lengths each pulse gets spread out over time as it travels. This is called **multimode dispersion**.

The spreading of individual pulses means that pulses that start out separated by a short time interval might overlap (Fig. 2.6.9) after they have travelled through a length of fibre. So the use of multimode fibres is restricted to short distances (such as inside buildings) or to less rapid streams of data (fewer pulses per second).

Fig. 2.6.9 Spreading and overlap of pulses.

## Example

An optical fibre is 1.60 km long. The refractive index of its core is 1.50. The greatest angle, $\alpha$, to the axis (Fig. 2.6.8) at which light can travel through the core without escaping is 14°. Calculate:

(a) The refractive index of the cladding.

(b) The difference in times of travel of light by the longest and shortest paths.

## Answer

(a) The critical angle at the core/cladding boundary is $90° - 14° = 76°$.

  So $1.50 \sin 76° = n_{clad} \sin 90°$  →  $n_{clad} = 1.46$

(b) Length of longest path $= \dfrac{AB}{AC} 1600 \text{ m} = \dfrac{1}{\cos 14°} 1600 \text{ m} = 1649 \text{ m}$

  So $t_{max} = \dfrac{1649 \text{ m}}{2.00 \times 10^8 \text{ m s}^{-1}} = 8.24 \text{ μs}$

  So $t_{max} - t_{min} = 8.24 \text{ μs} - 8.00 \text{ μs} = 0.24 \text{ μs}$

For transfer of rapid streams of data over long distances **monomode fibres** are used. They have very thin cores (a few wavelengths in diameter). In these fibres there are no zigzag paths; light can travel only parallel to the axis. So there is no multimode dispersion. [The reason is well beyond A-level!]

# Extra questions

1. Calculate the refractive index of the cladding in Quickfire 6.

2. The refractive index of sea water and ice is 1.33 and 1.31 respectively.
   (a) Calculate the critical angle for the boundary between these two materials.
   (b) State the circumstances in which TIR would occur at a boundary between these materials.

3. A multimode fibre has a core of refractive index 1.530 and a cladding of refractive index 1.520.
   (a) Determine the maximum angle between a light path and *the axis* of the fibre if light is to travel for a long distance through the fibre, accompanying your answer with a diagram.
   (b) Explain why it is an advantage for this angle to be small for transmission of data.

4. A laser beam is directed on to the end-face of a clear plastic rod of refractive index 1.35, surrounded by air.

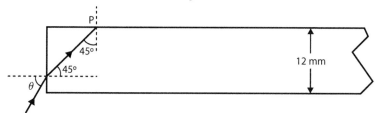

   (a) Calculate the value of $\theta$.
   (b) At **P**, 90% of the light power is refracted out of the glass and 10% is reflected. How far does the reflected light travel along the rod *from* **P** before the power drops to a millionth of the power of the beam incident on **P**? Show your working. [Consider successive reflections.]
   (c) The angle of incidence, $\theta$ is now decreased just enough for light to be **totally** internally reflected near P. Calculate the new value of $\theta$.

5. The speed of sound in air increases with temperature. The temperature of the atmosphere decreases with height. A sound wave is sent up from the ground at an angle of 45° to the horizontal. Draw a diagram to show how its direction changes. Briefly explain your answer.

6. Explain why the angles $\theta_W$ in Quickfire 3 and the fish tank example are the same (i.e. explain why the presence or absence of the glass is irrelevant for this calculation).

7. Add partial reflections to your answer to Quickfire 5. Comment on the strength of these.

# 2.7 Photons

## 2.7.1 The idea of a photon

Everything in nature seems to come in lumps or **quanta**. For example, ordinary matter is made of atoms, and electric charge comes in units of $e$. This lumpiness or *quantisation* was becoming fully accepted only in the early 1900s. In 1905 Einstein boldly suggested that light, too, was quantised. Light quanta are now called **photons**.

A photon, then, is a discrete packet of electromagnetic (e-m) radiation energy. The energy of a photon is given by

$$E_{phot} = hf$$

in which $f$ is the frequency of the e-m radiation and $h$ is a constant called the **Planck constant**. By experiment (see following sections), $h = 6.63 \times 10^{-34}$ J s.

Using $c = f\lambda$ (in which $c = 3.00 \times 10^8$ m s$^{-1}$) we can express the photon energy in terms of the *wavelength* of the radiation:

$$E_{phot} = \frac{hc}{\lambda}$$

### Example 1

The energy of a photon (one of a pair arising from the annihilation of an electron and a positron) is $8.19 \times 10^{-14}$ J. What wavelength of radiation does this represent?

**Answer**

Re-arranging the second equation and putting in values:

$$\lambda = \frac{hc}{E_{phot}} = \frac{6.63 \times 10^{-34} \text{ J s} \times 3.00 \times 10^8 \text{ m s}^{-1}}{8.19 \times 10^{-14} \text{ J}} = 2.43 \times 10^{-12} \text{ m}$$

This is in the $\gamma$-ray region of the e-m spectrum (Section 2.7.2); note that the wavelength of visible light is about 200 000 times greater!

### Example 2

A light-emitting diode produces 0.70 W of light of mean wavelength 630 nm. How many photons does it emit per second?

**Answer**

$$E_{phot} = \frac{hc}{\lambda} = \frac{6.63 \times 10^{-34} \text{ J s} \times 3.00 \times 10^8 \text{ m s}^{-1}}{630 \times 10^{-14} \text{ m}} = 3.16 \times 10^{-19} \text{ J}$$

$$\text{photon rate} = \frac{\text{energy per second}}{\text{energy per photon}} = \frac{0.7 \text{ J s}^{-1}}{3.16 \times 10^{-19} \text{ J photon}^{-1}}$$

$$= 2.2 \times 10^{18} \text{ photon s}^{-1}$$

---

>> **Pointer**

'Quanta' is a plural word. Its singular form is 'quantum'.

**Key Terms**

A **photon** is a discrete packet of electromagnetic (e-m) radiation energy.

**Photon energy**, $E_{phot} = hf$, in which $f$ is the frequency of the e-m radiation and $h$ is the **Planck constant**.

[$h = 6.63 \times 10^{-34}$ J s]

**quickfire**

① Calculate the energy of a photon of ultraviolet radiation of wavelength 100 nm.

>> **Pointer**

Some evidence for photons:

• The photoelectric effect (Sections 2.7.3–2.7.6) gives some of the most direct evidence.

• Atomic line spectra (Section 2.7.7) require photon model

• Light emitting diodes (LEDs) can be used (Section 2.7.12) to find a value for $h$, with the aid of a simple hypothesis involving photons.

# 2.7.2 The electromagnetic (e-m) spectrum

## (a) The visible spectrum

A diffraction grating (Section 2.5.7) splits 'white light' (e.g. sunshine) into a continuous spectrum of colours. In the first order (and higher orders) these colours of light emerge at different angles, showing that they have different wavelengths. The colours range from dark red, $\lambda \sim 700$ nm, through shades of orange, yellow, green and blue to violet ($\lambda \sim 400$ nm).

## (b) The whole e-m spectrum

Our eyes are sensitive to only a tiny slice of the whole spectrum of e-m waves. The other regions of the spectrum were discovered at different times, using different sources, and were given various names. Hence the overlap between X-rays and gamma rays in Fig. 2.7.1. For example, radiation of wavelength $1 \times 10^{-11}$ m could come from an X-ray tube *or* from certain radioactive nuclei.

Fig. 2.7.1 uses logarithmic scales: tenfold increases are represented by one scale division. Half a division is a factor of $\sqrt{10}$ (about 3).

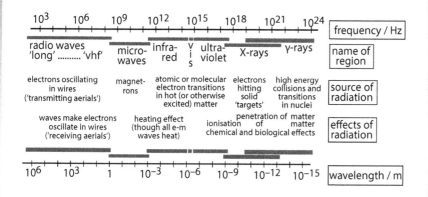

*Fig. 2.7.1 The electromagnetic spectrum*

### Example featuring the electron volt (eV)

In a certain X-ray tube, electrons emitted by a hot filament are accelerated towards a metal 'target' by a pd of $1.2 \times 10^5$ V. We say that each electron loses electrical PE of $1.2 \times 10^5$ **electron volt** ($1.2 \times 10^5$ eV), and therefore gains this amount of KE by the time it hits the target.

When one of these electrons hits the target, a photon may be emitted. The highest possible energy it could have is $1.2 \times 10^5$ eV. Calculate the wavelength of a photon of this energy.

## quicKpire

④ A laser emits photons of energy 1.27 eV. Calculate the wavelength of radiation emitted, and name the region of the e-m spectrum in which it lies.

**Answer**

$$E_{phot} = 1.2 \times 10^5 \text{ eV} = 1.2 \times 10^5 \text{ V} \times 1.60 \times 10^{-19} \text{ C} = 1.92 \times 10^{-14} \text{ J}.$$

$$\text{So } \lambda = \frac{hc}{E_{phot}} = \frac{6.63 \times 10^{-34} \text{ J s} \times 3.00 \times 10^8 \text{ m s}^{-1}}{1.5 \times 10^5 \text{ V} \times 1.60 \times 10^{-19} \text{ C}} = 1.0 \times 10^{-11} \text{ m}$$

# 2.7.3 The photoelectric effect

When electromagnetic radiation of high enough frequency falls on a metal surface, electrons are emitted from the surface.

For most metals, ultraviolet radiation is needed. For caesium, visible light (but not far red) will release electrons.

## Key Terms

**The photoelectric effect** is the emission of electrons from the surface of a metal due to e-m radiation of high enough frequency falling on it.

If you've seen the (rather striking) demonstration using a metal plate attached to a gold-leaf electroscope, check that you can recall the procedure! The demonstration we now describe uses a **vacuum photocell** (see diagram)...

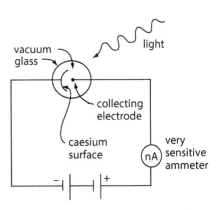

Fig. 2.7.2 Demonstrating the photoelectric effect

When light (of any colour but red) falls on the caesium surface, the very sensitive (nano-) ammeter registers a current.

Electrons emitted from the surface into the vacuum are attracted to the collecting electrode (made positive by the battery).

The electrons flow through the ammeter and the battery back to the caesium surface.

In fact there is a current is even if the battery is replaced by a piece of wire. This is because some electrons are emitted with enough KE to reach the collecting electrode by themselves.

## quicKpire

⑤ If the nano-ammeter in Fig. 2.7.2 reads 140 nA, calculate the number of electrons reaching the collecting electrode per second.

## 2.7.4 Measuring $E_{k\,max}$

In the circuit of Fig. 2.7.3 we increase the pd between the caesium surface and the collecting electrode, making the collecting electrode more negative, until the current *just* drops to zero.

At this point, the pd is called the stopping voltage, $V_{stop}$, because it stops all emitted electrons from reaching the collecting electrode. Those with the maximum KE *almost* reach the collecting electrode before being stopped, so gaining electrical PE of $eV_{stop}$, in exchange for their KE lost.

So $\qquad E_{k\,max} = eV_{stop}$

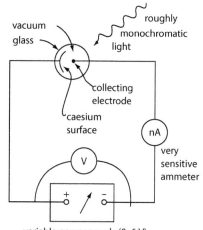

Fig. 2.7.3 Determining $E_{k\,max}$ of emitted electrons

**Example**

If $V_{stop} = 1.7$ V, then $E_{k\,max} = 1.60 \times 10^{-19}$ C $\times$ 1.7 V $= 2.7 \times 10^{-19}$ J

Alternatively we can say (see Section 2.7.2) that $E_{k\,max} = 1.7$ eV.

### quickpire

(6) When a caesium surface is illuminated with light of a certain frequency, the stopping voltage for emitted electrons is 0.35 V. Determine their maximum KE in eV and in J.

## 2.7.5 Plotting $E_{k\,max}$ against frequency, $f$

We find $E_{k\,max}$, as just described, for three or four known frequencies, $f$, of light falling in turn on a metal surface. Each could come from a low-power laser. The wavelength could be measured with a diffraction grating, and the frequency calculated using $f = c/\lambda$.

We plot $E_{k\,max}$ against $f$. The points lie on a straight line (Fig. 2.7.4).

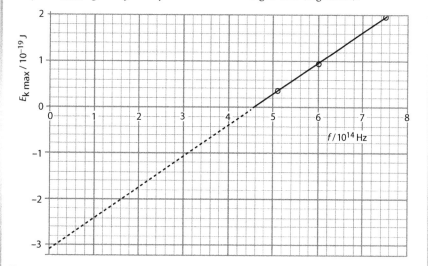

Fig. 2.7.4 $E_{k\,max}$ against $f$ for a caesium surface

### Key Term

**Key term**

The **work function**, $\phi$, of a metal is the minimum energy needed to eject an electron from its surface.

### quickpire

(7) For the point on the Fig. 2.7.4 graph at $6.0 \times 10^{14}$ Hz calculate the maximum KE of an emitted electron in $eV$.

### ≫ Pointer

Don't confuse the work function of a *surface* with the ionisation energy of an atom.

# 2.7.6 Einstein's photoelectric equation

Although the free electrons in a metal have no ties to *particular* atoms, there are forces holding them to the lattice of atoms (strictly, ions) as a whole. To escape from the metal an electron has to do work against these forces. Some electrons have to do less work than others, but there is a certain minimum amount needed called the **work function**, $\phi$.

The key idea behind Einstein's photoelectric equation is that any electron that leaves the surface is ejected by a single photon. Photons don't co-operate in the ejection.

Suppose a photon gives its energy, $hf$, to an electron, which is able to escape. The minimum energy used in escaping is $\phi$, leaving $hf - \phi$ as the maximum kinetic energy ($E_{k\,max}$) that the escaped electron can have.

Thus $\quad E_{k\,max} = hf - \phi$

## (a) How the graph of $E_{k\,max}$ against $f$ fits the equation $E_{k\,max} = hf - \phi$

Comparing the equation with $y = mx + c$ (see Pointer), it is clear that a graph of $E_{k\,max}$ against $f$ should be straight, with positive gradient and negative intercept. This is confirmed by experiment (Fig. 2.7.4). Note that:

Graph gradient = $h$; graph intercept = $-\phi$

## b) Minimum photon energy to release electrons

For *any* electrons to be ejected, the photon energy, $hf$, must be at least as big as the work function, $\phi$. We write:

$$hf_{thresh} = \phi \quad \text{that is} \quad f_{thresh} = \frac{hc}{\lambda}$$

in which $f_{thresh}$ is the **photoelectric threshold frequency,** of the metal.

### Example

Light of wavelength 480 nm liberates electrons with a maximum kinetic energy of 0.309 eV from a sodium surface. Calculate the work function of sodium and the maximum wavelength, $\lambda_{max}$, of light that will eject electrons from a sodium surface.

### Answer

$E_{k\,max}$ needs to be expressed in J. We shall also need $f = c/\lambda$.

$$E_{k\,max} = hf - \phi \quad \text{that is} \quad \phi = \frac{hc}{\lambda} - E_{k\,max}$$

$$\text{So } \phi = \frac{6.63 \times 10^{-34} \text{ J s} \times 3.00 \times 10^8 \text{ m s}^{-1}}{480 \times 10^{-9} \text{ m}} - 0.309 \times 1.60 \times 10^{-19} \text{ J}$$

This gives $\phi = 3.65 \times 10^{-19}$ J [$= 2.28$ eV]

and $f_{\text{thresh}} = \dfrac{\phi}{h} = \dfrac{3.65 \times 10^{-19} \text{ J}}{6.63 \times 10^{-34} \text{ J s}} = 5.51 \times 10^{14}$ Hz

So $\lambda_{\text{max}} = \dfrac{3.00 \times 10^{8} \text{ m s}^{-1}}{5.51 \times 10^{14} \text{ Hz}} = 545$ nm

## quickfire

⑩ Read the photoelectric threshold frequency for caesium from Fig. 2.7.4. Calculate the corresponding wavelength and state the colour of light.

## (c) Effect of light intensity on emitted electrons

The **intensity** of light on a surface is the energy of electromagnetic radiation that falls per m², per second on that surface.

Suppose we increase the intensity of light on an emitting surface, without changing the frequency of the light. This will increase the number of photons falling per second on the surface, but won't affect the energy of the individual photons. We can therefore predict that:

## quickfire

⑪ The work function of potassium is 2.3 eV. Calculate the frequency of e-m radiation needed to eject electrons of maximum KE 3.5 eV from a potassium surface and state its region of the e-m spectrum.

- More photons arriving per second will eject more electrons per second. This shows up as an increase in the current, which can be confirmed using the vacuum photocell in the circuit of Fig. 2.7.2.

- $E_{\text{k max}}$ is unaltered. Recall that photons don't co-operate in ejecting electrons. This is easily confirmed: $V_{\text{stop}}$ is unchanged.

On a pure 'wave picture' of light, without photons, the last result can't be explained simply. Neither can the existence of a threshold frequency.

# 2.7.7 Atomic line emission spectra

## (a) Producing and observing the spectra

Light with a line spectrum is emitted from excited unattached atoms, for example those in the sodium vapour in a street lamp, or those that dissociate from salt crystals sprinkled into a hot flame.

In the street lamp the atoms are **excited** (have their energy levels raised) by hits from 'missile' electrons. In the flame the atoms are excited by random hits from fast-moving molecules.

To examine the spectrum of the light emitted we use a diffraction grating (Section 2.5.7). We find (repeated in each order) a **line emission spectrum** of sharp, bright, coloured lines, corresponding to (almost) single wavelengths.

≫ *Pointer*
'Spectra' is a plural word.
'Spectrum' is its singular.

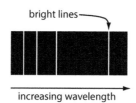

bright lines⟶

*Fig. 2.7.5 Line emission spectrum*　　increasing wavelength⟶

### Pointer

Don't be put off by the negative energies in Fig. 2.7.7. They arise simply because it's conventional to assign zero energy to the ionisation level in the H atom.

### Grade boost

The longer the transition arrow the shorter the photon wavelength!

#### quickfire

⑫ Use Fig. 2.7.7 to determine the ionisation energy of an H atom. Give the answer in joules.

#### quickfire

⑬ Calculate the wavelengths emitted in transitions (b) and (c) in Fig. 2.7.7

#### quickfire

⑭ In which region of the e-m spectrum is the radiation from transition (c)? How can you now be sure that the radiation from *all* transitions to the ground state in a hydrogen atom will be in this region?

## (b) How a line emission spectrum arises

An atom can exist only in certain states, each with a definite energy. Fig. 2.7.6 shows these 'energy levels' for hydrogen, the simplest atom, having only one electron. The atom will be in its ground state (lowest energy) unless hit, for example by other atoms.

The lower the energy of the state, the closer to the nucleus the electron is likely to be found. The **ionisation energy** is the energy needed to *remove* an electron from an atom in its ground state.

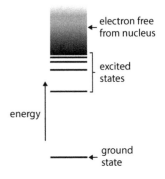

Fig. 2.7.6 Hydrogen atom energy levels

Suppose the atom (or, if you prefer, its electron) has been excited to the second excited level (see Fig. 2.7.7). It soon returns to the (stable) ground state, either in one transition, (a), or in two, (b) and (c).

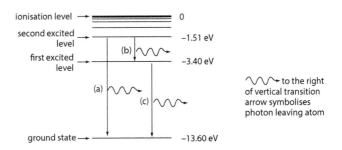

Fig. 2.7.7 Hydrogen atomic transitions

In any of these 'downward' transitions, the energy lost by the atom is **spontaneously emitted** as a single photon. So if the atom goes between levels with energies $E_U$ (upper) and $E_L$ (lower),

$$hf = E_U - E_L \quad \text{so} \quad f = \frac{E_U - E_L}{h}$$

e.g. in transition (a):     $E_U - E_L = -1.51 \text{ eV} - (-13.60 \text{ eV}) = 12.09 \text{ eV}$

So, converting 12.09 eV to J

$$f = \frac{12.09 \times 1.60 \times 10^{-19} \text{ J}}{6.63 \times 10^{-34} \text{ J s}} = 2.92 \times 10^{15} \text{ Hz (in the ultraviolet)}$$

# 2.7.8 Atomic line absorption spectra

## (a) Producing and observing the spectra

A line absorption spectrum arises when a beam of e-m radiation, with a continuous range of frequencies, is sent through an 'atmosphere' containing unattached atoms.

When the light that emerges is passed through a diffraction grating, dark lines are seen to cross the continuous spectrum.

The dark lines are at the exact wavelengths of some of the bright lines in the atoms' emission spectrum!

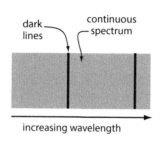

Fig. 2.7.8 Line absorption spectrum

>> **Pointer**
No equations for frequencies of absorbed photons are given. This is because exactly the same equations apply as for *emitted* photons!

**quickfire**

⑮ Suppose we find a dark line of wavelength corresponding to transition (b) (Fig. 2.7.9) in the absorption spectrum of a star. What can we conclude about the temperature of its atmosphere?

## (b) How a line absorption spectrum arises

An atom can make a transition from a lower energy level ($E_L$) to a higher energy ($E_U$) by absorbing a photon. Only a photon with energy ($E_U - E_L$) can be absorbed. A photon with a different energy will continue on its way – ignored by the atom!

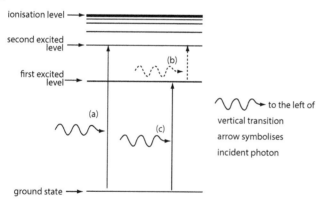

Fig. 2.7.9 Some absorption transitions for hydrogen atom

In Fig. 2.7.9, (a) and (c) represent two possible absorption events for a hydrogen atom. Event (b) won't happen at room temperature, as the atoms aren't being buffeted enough (by other atoms) for any of them to be at even the first excited level, so they can't be promoted from it! We say that, at room temperature, the excited levels are not 'populated'.

>> **Pointer**
An atom that has been excited by absorbing a photon will spontaneously emit a photon, often of the same frequency. Won't this 'make up for' the absorbed photon? No, because the atoms emit in random directions, so the original beam is still depleted in the absorbed frequencies.

**Grade boost**
Check that you can explain how the dark lines arise in an absorption spectrum. Don't forget about 're-emission'.

## (c) Identifying atoms from their spectra

Atoms of different elements have different energy levels, and hence different line spectra, so we can identify atoms from their spectra – even for atoms light years away! See Section 1.6.

# Grade boost

The formula $p = \dfrac{h}{\lambda}$ for photon momentum is quite different from the familiar $p = mv$, but the units must agree. Check that those of $h/\lambda$ can be written as kg m s⁻¹.

# Pointer

Because a photon's energy can be written as
$$E_{phot} = \frac{hc}{\lambda}$$
we now have
$$E_{phot} = cp_{phot}$$
This is reassuring because, even before photons were thought of, e-m radiation was known to have both energy and momentum. For the radiation crossing a given area in a given time it had been shown that:
$$\text{energy} = c \times \text{momentum}$$

## quickfire

⑯ The excited nucleus of a $^{24}_{11}$Na atom emits a γ-ray photon of wavelength $4.50 \times 10^{-13}$ m. Calculate:
(a) The photon's momentum.
(b) The recoil speed of the atom. Its mass is $3.98 \times 10^{-26}$ kg.

# 2.7.9 Photon momentum

A photon is a discrete packet of e-m wave energy. It has no mass, but it does have momentum of a definite magnitude, $p$, given by the equation $p = \dfrac{h}{\lambda}$ in which $h$ is the Planck constant and $\lambda$ is the wavelength.

Momentum is a vector quantity, and the direction of a photon's momentum is simply the direction of travel of the e-m radiation.

Photons can be absorbed and emitted and can take part in collisions. The principle of conservation of momentum applies every time.

### Example

A photon of wavelength $9.52 \times 10^{-8}$ m is absorbed by a hydrogen atom of mass $1.67 \times 10^{-27}$ kg that is initially stationary. Calculate the speed the atom acquires. (Its mass change is negligible.)

### Answer

Total momentum before = total momentum after, so $\dfrac{h}{\lambda} + 0 = mv$

$$\text{So } v = \frac{h}{m\lambda} = \frac{6.63 \times 10^{-34} \text{ J s}}{1.67 \times 10^{-27} \text{ kg} \times 9.52 \times 10^{-8} \text{ m}} = 4.2 \text{ m s}^{-1}$$

## Calculating the force of light on a surface

The atom in the last example acquires the photon's momentum, and experiences a force as the absorption takes place.

If a stream of photons is absorbed by a body, there will be a succession of such hits. We can calculate the *mean* force on the body while the photons are impinging using

$$\text{Mean force} = \frac{\text{photon momentum absorbed}}{\text{time taken}}$$

$$= \frac{\text{number of photons absorbed} \times \text{momentum per photon}}{\text{time taken}}$$

If the photons bounce back, or as many are re-emitted as absorbed, there's twice the momentum change – and twice the mean force.

### Example

A perfectly reflective 'solar sail' attached to a spacecraft has an area of 25 m². Calculate the force exerted on it by sunshine of intensity 1.2 kW m⁻² falling normally on it. You may assume that the light behaves as if it were monochromatic, of wavelength 550 nm.

### Answer by piecemeal arithmetical calculation

$$\text{Energy of photon} = \frac{hc}{\lambda} = \frac{6.63 \times 10^{-34} \text{ J s} \times 3.00 \times 10^{8} \text{ m s}^{-1}}{550 \times 10^{-9} \text{ m}} = 3.62 \times 10^{-19} \text{ J}$$

$$\text{So} \quad \text{number of photons per second} = \frac{1200 \text{ W m}^{-2} \times 25 \text{ m}^2}{3.62 \times 10^{-19} \text{ J}} = 8.29 \times 10^{22} \text{ s}^{-1}$$

But momentum of photon $= \dfrac{h}{\lambda} = \dfrac{6.63 \times 10^{-34} \text{ J s}}{550 \times 10^{-9} \text{ m}} = 1.21 \times 10^{-27} \text{ N s}$

So   mean force $= 2 \times \dfrac{8.29 \times 10^{22} \text{ s}^{-1} \times 1.21 \times 10^{-27} \text{ J}}{1 \text{ s}} = 0.20 \text{ mN}$

**Answer using symbols up to the last moment**

($I$ = light intensity, $A$ = surface area.)

Energy of photon $= \dfrac{hc}{\lambda}$

So   number of photons per second $= IA \div \dfrac{hc}{\lambda} = \dfrac{IA\lambda}{hc}$

But momentum of photon $= \dfrac{h}{\lambda}$

So   mean force $= 2 \times \dfrac{IA\lambda}{hc} \times \dfrac{h}{\lambda} = \dfrac{2IA}{c} = \dfrac{2 \times 1200 \text{ W m}^{-2} \times 25 \text{ m}^2}{3.00 \times 10^8 \text{ m s}^{-1}} = 0.20 \text{ mN}$

Given a few years, that spacecraft could reach quite a speed!

In fact the force of e-m radiation does have to be allowed for when planning space voyages. It also visibly deflects the tails of comets. In very large, hot stars the force of radiation travelling outwards prevents the star from collapsing under gravitational forces.

## 2.7.10 Particles with mass have wave properties

The idea of photons dates from 1905, when Einstein suggested that light interacted with matter as discrete packets of energy. This is sometimes described as light having particle properties (in addition to its wave properties). Almost twenty years later, Louis de Broglie suggested that things with mass, previously thought of as particles (such as electrons, protons, atoms, marbles...), also had wave properties. For a particle with momentum of magnitude $p$, the wavelength, he argued, would be $\lambda = \dfrac{h}{p}$.

This is exactly the same relationship that applies to a photon, though for a particle we have in addition $p = mv$ (provided that the particle's speed, $v$, is much less than $c$).

**Example**

Calculate the wavelength of an electron that has been accelerated from rest through a pd of 2.5 kV.

**Answer**

The electron will possess KE, $E_k$, of 2.5 keV $= 2500 \times 1.60 \times 10^{-19}$ J

But $E_k = \frac{1}{2} m_e v^2$   and   $p = m_e v$

So, eliminating $v$, $E_k = \dfrac{p^2}{2m_e}$ and therefore $p = \sqrt{2m_e E_k}$

So $\lambda = \dfrac{h}{p} = \dfrac{6.63 \times 10^{-34} \text{ J s}}{\sqrt{2 \times 9.11 \times 10^{-31} \text{ kg} \times 2500 \times 1.60 \times 10^{-19} \text{ J}}} = 2.46 \times 10^{-11} \text{ m}$

This calls out to be tested experimentally...

**Grade boost**

In the example, note how working in symbols as long as possible saved calculation, writing – and *time*.

In this case one also gains insight. $\lambda$ turns out to be irrelevant. For a given light intensity a larger $\lambda$ would give photons of less momentum, but, compensatingly more of them. We didn't *even* need to assume the light to be monochromatic.

**quickfire**

⑰ It is proposed to levitate (that is prevent from falling) a leaf of mass 0.14 g, by shining light of wavelength 630 nm from a laser on to its underside. Calculate the required power of the laser beam assuming the leaf to be perfectly reflecting. What, in practice, would happen to the leaf?

**quickfire**

⑱ Determine the speed, $v$, of the electron in the example. It should be much less than $c$, justifying the use of the ordinary formulae for $p$ and $E_k$.

# 2.7.11 Demonstrating electron diffraction with crystals

Wavelengths of the order of magnitude just calculated would be far too small to produce distinguishable beams from a conventional diffraction grating, even with its $d$ as low as $1 \times 10^{-7}$ m. But the atoms in a crystal are arranged regularly with a typical separation of $2 \times 10^{-10}$ m, allowing crystals to be used as 'natural' diffraction gratings for wavelengths down to ~$10^{-12}$ m. [They were first used in this way with X-rays in 1912.]

In Fig. 2.7.10, many very small graphite crystals are used as gratings. The apparatus is simpler than it might look. We start with the electron gun. The current through the tungsten filament makes it so hot that electrons are continually being knocked out of it. These electrons are accelerated towards the (metal) 'anode' by the high voltage placed between the filament and the anode (with the anode positive). Some of these electrons shoot through the small hole in the anode, forming a beam of fast electrons directed at the graphite crystals.

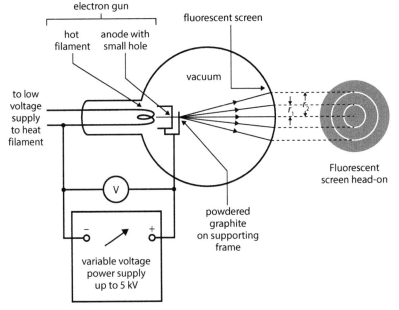

Fig. 2.7.10 Demonstrating electron diffraction

The fluorescent screen emits photons from points where electrons hit it (with enough KE). So the bright rings we see on the screen are where most electrons hit. Beams of electrons must therefore be leaving the crystals at specific angles – like waves from diffraction gratings!

# 2.7.12 Specified practical work – using LEDs to obtain a rough value for the Planck constant

The photoelectric equation is probably the simplest soundly-based, testable consequence of the photon aspect of light. However, for experimental results that give a reliable value for $h$, the electron-emitting surface has to be ultra-pure, so good vacuum photocells are expensive!

We give below a quite different experiment using light-emitting diodes (LEDs). By making an assumption that is known not to be perfectly accurate we can determine $h$ roughly.

## (a) An approximate relationship

When a small pd is applied one way round to the LED, there is a current, and light is emitted. The simplest assumption we can make – unlikely to be exactly true – is that, when a photon is emitted:

Energy of an emitted photon = Electrical energy lost by an electron

$$\text{So} \qquad \frac{hc}{\lambda} = eV_s$$

in which $V_s$ is the minimum pd needed across the LED to produce light, its so-called 'striking voltage'. The idea is to take readings of $V_s$ and $\lambda$ for diodes that give light of different colours.

## (b) The experiment

(a) Increase the power supply voltage from zero until the LED, viewed down a black paper tube, *just* starts to glow. Read $V_S$ from the voltmeter.

If the LED doesn't glow, check that it's connected the right way round!

Repeat with a few diodes that give different colours.

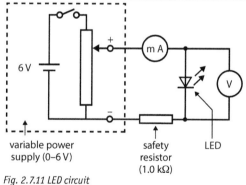

variable power supply (0–6 V)    safety resistor (1.0 kΩ)    LED

*Fig. 2.7.11 LED circuit*

(b) Use a diffraction grating (Section 2.5.7) to determine the peak wavelength, $\lambda$, emitted by each LED, You will need to turn up the power supply so that the LEDs give out enough light. Don't exceed the maximum allowed current. [Alternatively, use the manufacturer's Data Booklet to give you the mean emission wavelength for the LED.]

(c) Plot $V_S$ against $\frac{1}{\lambda}$. Draw the straight line of best fit, not necessarily through the origin. See Quickfire 23 and the final Pointer.

>> **Pointer**

Inside the transparent case of an LED is a piece of semiconductor engineered to have a different composition either side of a so-called 'junction'. The wires connect either side. An electron crossing the junction region can drop from one energy level to another, emitting a photon.

**quicKpire**

21 What is the name of the circuit labelled 'power supply' in Fig. 2.7.11?

>> **Pointer**

It's fine to use a low voltage power supply 'off the shelf'. It must give 'smoothed' dc.

**quicKpire**

22 Use the equation to put these colours of LED in predicted order of striking voltage (lowest first): green, blue, red, yellow.

**quicKpire**

23 What does the gradient, $m$, of the graph of $V_S$ against $\frac{1}{\lambda}$ represent? How is the value of $h$ determined?

>> **Pointer**

The graph points 'should' lie on a straight line through the origin. If the fit is not good, consider why it might not be, and whether any patterns are apparent.

# Extra questions

1. State which of the three terms in Einstein's photoelectric equation depends on:
   (a) The surface on which the light falls, but not on the light itself.
   (b) The light (or ultraviolet) falling on the surface, but not on the surface itself.

2. Light of frequencies given below falls on a caesium surface of work function 2.10 eV. In each case calculate the maximum kinetic energy of the emitted electrons. Explain your answer. If emission does not take place, explain, in terms of *energy*, why not.
   (a) Violet light of frequency $7.0 \times 10^{14}$ Hz.
   (b) Green light of frequency $5.7 \times 10^{14}$ Hz.
   (c) A mixture of violet light of frequency $7.0 \times 10^{14}$ Hz and green light of frequency $5.7 \times 10^{14}$ Hz.
   (d) Red light of frequency $4.5 \times 10^{14}$ Hz.

3. In an energy level diagram of a hydrogen atom, the lowest three levels have energies of $-2.18 \times 10^{-18}$ J, $-0.54 \times 10^{-18}$ J and $-0.24 \times 10^{-18}$ J, when the ionisation level is assigned an energy of zero.
   (a) State the ionisation energy of a hydrogen atom, in eV.
   (b) A continuous spectrum of ultraviolet and visible radiation is shone through a tube containing hydrogen atoms. The radiation that emerges is found to have a number of missing wavelengths ('dark lines').
      (i) Explain how the dark lines arise in the ultraviolet region.
      (ii) Calculate the wavelength of one of these dark lines.
      (iii) Discuss whether or not you would expect to find a dark line crossing the visible region due to transitions between the first and second excited states.

4. (a) Determine the energy and the momentum of a photon of light of wavelength 630 nm.
   (b) Derive an equation for the force exerted on a reflecting surface by a laser beam of power $P$ directed normally at the surface.

5. (a) Calculate the kinetic energy of a proton with a wavelength of $3.96 \times 10^{-14}$ m, giving your answer in eV.
   (b) Explain, without using a calculator, whether an electron of the same kinetic energy would have a larger or smaller wavelength.

# 2.8 Lasers

## 2.8.1 Properties and origin of laser light

A **laser** produces a beam of coherent light: very nearly monochromatic (single frequency), with wavefronts extending across the width of the beam, as though coming from a point source. This makes laser light useful for precision measurement, and for carrying encoded data. Some lasers produce intense narrow beams, useful for surgery – or welding!

Laser light is generated by **stimulated emission**, described below. This process occurs in the so-called **amplifying medium** of the laser. We shall assume it to take place within individual atoms of the medium.

## 2.8.2 Stimulated emission of radiation

Fig. 2.8.1 shows, symbolically, three processes involving electron energy levels $E_U$ and $E_L$, and photons of energy $(E_U - E_L)$. [$E_L$ might be the ground state, and $E_U$, the first excited state.]

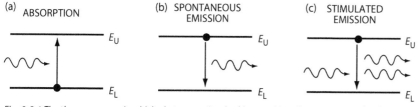

Fig. 2.8.1 The three processes in which photons are involved in transitions between energy levels

The first two processes were discussed, in Sections 2.7.7–2.7.8.

In **stimulated emission**, a photon with energy $(E_U - E_L)$, passing near the electron at level $E_U$, causes it to 'drop' to $E_L$, emitting a photon, of energy $(E_U - E_L)$. So we now have *two* photons of the same frequency – the beginnings of **light amplification**. The emitted photon is *in phase* with the stimulating photon, polarised in the same direction and travelling in the same direction.

## 2.8.3 Populations

For laser action ('lasing') this 'cloning' must happen repeatedly, creating a photon avalanche. But without special measures, those photons that interact with electrons will just be absorbed, rather than causing stimulated emission. This is because, at normal temperatures, almost all electrons will be in the ground state; we say that the ground state is *fully populated* and excited levels are *empty* – as symbolised in Fig. 2.8.2 (a).

### Key Terms

**Laser** is an acronym for Light Amplification by Stimulated Emission of Radiation.

**Stimulated emission** is the emission of a photon from an excited atom, triggered by a passing photon of energy equal to the energy gap between the excited state and a state of lower energy.

The **amplifying medium** is the (transparent) material in which stimulated emission takes place.

 *Grade boost*

Check that you can state four things in common between the stimulating and emitted photons. The 'stacked wiggles' in Fig. 2.8.1 (c) symbolise these things; they don't *depict* photons in space.

> **Pointer**
> Note the little blobs on the energy levels in Figs 2.8.1 and 2.8.2. A blob represents an electron with that energy.

> **Pointer**
> In Fig. 2.8.2, identical energy levels of different atoms have been 'run together'.

> **Pointer**
> One method of pumping, called *optical pumping*, is to shine a very bright light containing photons of energy ($E_P - E_G$) at the atoms.

### quickfire

① The first laser producing visible light (deep red at 694 nm) was built by Theodore Maiman in 1960. It was a three-level system pumped with light from a xenon 'flash tube'.

Calculate the energy of the upper lasing transition level above the level of the ground state, in J and eV.

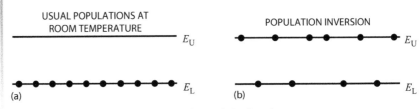

Fig. 2.8.2 Room temperature populations and a population inversion

To have more stimulated emission events (each *adding* a photon) than absorption events (each *taking away* a photon) we must bring about a **population inversion**: more electrons in the excited state than the ground state (Fig. 2.8.2 (b)).

## 2.8.4 A 2-level laser system?

How do we bring about the necessary population inversion?

Raising the atoms' temperature would increase the kinetic energy with which they randomly collide, and would promote some electrons to excited states. Never, though, to the extent of a population *inversion*.

Suppose, instead, that we shine on to the atoms a bright light with photon energy ($E_U - E_L$), hoping to promote electrons from the lower level (in this case the ground state) to the one directly above by photon absorption. If we manage to promote half the electrons to the upper level, that's as far as we can go, because at that stage as many photons will cause stimulated emission as absorption – stalemate.

We conclude that, in general, we cannot achieve a population inversion in a simple 2-level system. We can, though, in a 3- or 4-level system.

## 2.8.5 Workable laser systems

### (a) A 3-level laser system

- We **pump** energy into the system to promote electrons from the ground state, G, to a 'top' level, P. See second Pointer. For the required population inversion we have to pump fast enough to keep level G less than half full.

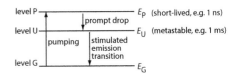

Fig. 2.8.3 A 3-level laser system

- The system has to be chosen so that level P is very short-lived, that is electrons spend only a very short time here before making spontaneous transitions to level $E_u$. This is so that P doesn't become full, preventing further pumping.

- Level U must be **metastable**: that is electrons spend, on average, a long time (even milliseconds!) here before spontaneously falling. In this way, given a high enough pumping rate, level U can maintain a higher population than level G. Stimulated emission is then favoured over absorption so, for photons of energy $(E_U - E_G)$.

## (b) A 4-level laser system

The key difference from a 3-level system is that the lower level, L, of the stimulated emission transition is *above* the ground state. This ensures that (at ordinary temperatures) it is self-emptying, by spontaneous transition to the ground state. The advantage is that we don't

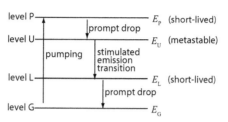

Fig. 2.8.4 A 4-level laser system

need such a ferocious rate of pumping (from G to P) in order to achieve population inversion between U and L, as L should always be nearly empty. (L needs to be short-lived.)

# 2.8.6 The laser itself

The amplifying medium is contained in the so-called **laser cavity** between two mirrors. The medium is pumped to maintain a population inversion. Photons of energy $(E_U - E_L)$. will arise by spontaneous emission within the medium. Any such photon can produce a clone by stimulated emission, so 1 photon becomes 2, 2 may become 4 and so on. This exponential increase is significant for photons travelling parallel to the axis of the laser because (by repeated reflections) they can traverse the medium many times, before escaping to form the beam.

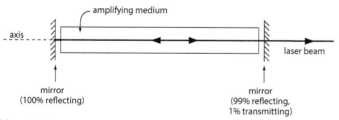

Fig. 2.8.5 Laser structure

The exponential increase in photon number doesn't go on forever. It is limited by the escape of photons through the not-quite-100%-reflecting mirror to form the beam, the absorption of photons, and the finite pumping rate.

## quickfire

② If, in a 4-level system,
$E_P = 2.72 \times 10^{-19}$ J
$E_U = 2.21 \times 10^{-19}$ J
$E_L = 0.30 \times 10^{-19}$ J
$E_G = 0$,
Calculate:

(a) The wavelength of light emitted by stimulated emission.

(b) The pumping energy needed per electron.

(c) The maximum possible efficiency of the system.

## quickfire

③ Light traversing the laser cavity in both directions results in a stationary wave (Section 2.5.8). There is a node at the 100% reflecting mirror and an approximate node at the other mirror, as well as intermediate nodes.

(a) Treating the stationary wave like that on a taut string, show that the possible wavelengths are given by $\lambda = 2L/n$, where $L$ is the cavity length and $n$ is a whole number.

(b) Hence show that a stationary wave of wavelength 820.0 nm can exist in a laser cavity of length 0.2050 mm, but a wave of wavelength 821.0 nm can't.

(c) Determine the next highest wavelength of stationary wave above 820 nm that can exist in the cavity.

## 2.8.7 Semiconductor (diode) lasers

The 'conventional' lasers we have been considering are very inefficient: less than 1% of the pumping energy emerges in the form of laser light. Much of the pumping energy fails to raise electron levels and finishes up as random thermal energy. Even for *successful* pumping events, $(E_P - E_G) > (E_U - E_L)$.

Most semiconductor lasers are modified diodes. See Section 2.7.12 (Pointer). Pumping, by means of an electric current, produces a population inversion in the amplifying medium, a thin layer of material (usually between *two* junctions). The mirrors are opposing faces of the chip itself, one of which is coated to assist reflection, while the other may let up to 60% of the light escape to form the beam. Their efficiency is generally far greater than that of conventional lasers. Semiconductor lasers are also much smaller and cheaper, though they are generally not capable of producing high power beams, nor beams that spread very little.

Uses of semiconductor lasers include 'scanning' CDs, DVDs, barcodes, and, in a laser printer, pages of text or pictures. Semiconductor lasers are also used for sending light, carrying digitally encoded data, into optical fibres.

# Extra questions

1. A 4-state gas laser is constructed with the following characteristics:
   - Pumped state $1.6 \times 10^{-18}$ J above the ground state.
   - The pumped state rapidly decays by collision and the loss of 4.8 eV.
   - The laser emits radiation of wavelength 500 nm.

   Draw and label an energy level diagram for this laser, giving energies in both eV and aJ.

2. A simplified energy level diagram for the amplifying medium of a 3-level laser is given:

   level P  ————————————————

   level U  ————————————————  $2.10 \times 10^{-19}$ J

   level O  ————————————————  0
   (ground state)

   (a) Suppose that the laser is at room temperature and that it is **not being pumped**.
       (i)  Compare the (electron) populations of the three levels.
       (ii) A photon of energy $2.10 \times 10^{-19}$ J in the laser cavity interacts with the amplifying medium. Name the process involved and explain briefly what happens.

   (b) The laser is now pumped, to create a population inversion between levels U and O.
       (i)   Use the diagram to help explain what is meant by a population inversion and how it is achieved.
       (ii)  Explain how light amplification takes place.
       (iii) Calculate the wavelength of the radiation emitted by stimulated emission.

# 3 Practical and data-handling skills

The AS qualification includes the indirect assessment of practical skills. Students who take the full A level in year 13 will be assessed on these skills directly, including on year 12 practical techniques. Examination papers also include questions on practical work and data handling. This chapter comprises a brief review of the skills involved.

## 3.1 Making and recording measurements

### 3.1.1 Resolution

**Key Term**

**Resolution** (of an instrument) = the smallest measurable change it can record.

(a) Digital instrument

This is 1 in the least significant figure in the display. For example, look at the display of a multimeter set on the 200 mA range:

The resolution is 0.1 mA.

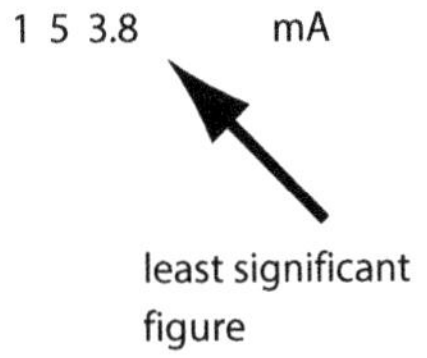

(b) Analogue instrument

Take the resolution as the interval between the smallest graduations, e.g. 1 mm for a metre rule. On the voltmeter scale in Fig. 3.2 the resolution is 0.1 V.

1 5 3.8    mA

least significant figure

*Fig. 3.1 Digital display*

*Fig. 3.2 Analogue display*

**» Pointer**

Examples of zero error
- Micrometers and vernier callipers often have a zero error, which may be positive or negative. To find this, close the jaws gently and note the reading.
- The ends of metre rules are often damaged – read from the 10.0 cm mark and subtract 10.0 cm from the reading.

### 3.1.2 Zero error

Always check that a measuring instrument reads 0 for a zero-sized measurement. If there is a non-zero least reading, you will need to allow for this.

**Example**

Resistance meter leads have resistance (typically around 0.4 Ω). Note the readings when the leads are touched together and subtract this from all subsequent readings on the same range.

### 3.1.3 How many significant figures (s.f.) do I write down?

The general rule is to use the resolution of the instrument to the full, e.g. when using a metre rule, write down the reading to the nearest mm, e.g. 115 mm, 65.0 cm (not just 65 cm).

If you depart from this rule, you should draw attention to it and say why.

**quickfire**

① What is the resolution of a protractor?

# 3.2 Displaying data

## 3.2.1 Tables – marking points

Over-arching heading

| Length / cm | Time for 20 oscillations / s | | | | Period / s |
|---|---|---|---|---|---|
| | Reading 1 | Reading 2 | Reading 3 | Mean | |
| 20.0 | 17.85 | 18.02 | 17.99 | 17.95 | 0.898 |
| 30.0 | 22.11 | 21.96 | 21.87 | 21.98 | 1.099 |
| 40.0 | | | | | |

Headings with units

Systematic presentation

Readings to the resolution of the instrument

Calculated data to be consistent

quicKpire

② An ohm-meter leads have a resistance of 0.3 $\Omega$. The reading with a length of wire is 17.6 $\Omega$. What is the wire's resistance?

## 3.2.2 Significant figures

This is slightly tricky because there are several factors to consider when deciding how many figures to claim in a quantity you write down. If you have an uncertainty estimate, then use that as the evidence – see Section 3.5. If not, use the figures in the data.

When multiplying and dividing, express the result to the same number of s.f. as in the least precise of the numbers used in the calculation.

### Example

A piece of aluminium sheet has dimensions 1.2 mm × 5.65 cm × 2.3 cm. Calculate the volume.

### Answer

If we work in cm:     Volume = 0.12 × 5.65 × 2.3 = 1.5594 cm³.

Two of the pieces of data have only 2 s.f., so we express the answer as 1.6 cm³.

If we worked in mm:     Volume = 1.2 × 565 × 23 = 1559.4 mm³, which we express as 1600 mm³ (2.s.f.).

Less commonly, we sometimes need to add or subtract data. In this case, we need to be careful of the number of decimal places in the answer.

### Example

Calculate the total mass of 3 people whose individual masses are 67 kg, 58.6 kg and 70 kg.

### Answer

Adding the numbers together gives 195.6 kg. If we assume that the 70 kg figure was given to the nearest kg (and not to the nearest 10 kg), then the answer for the total mass should be given as 196 kg.

quicKpire

③ What should the student have written?
  (a) 'The length of the A4 paper sheet = 30 cm' (measured using a metre rule).
  (b) 'Current = 7.5' (measured using a digital ammeter on mA scale, resolution 0.1 mA).
  (c) 'Mean value of resistance = 6.425 $\Omega$' (Calculated from values 6.3, 6.5, 6.4, 6.5 $\Omega$).

## 3.3 Graphs

### 3.3.1 Plotting graphs

**>> Pointer**

See Section 3.5 for graph plotting with error bars.

**>> Pointer**

Feel free to turn the grid on its side if that makes better use of the whole grid.

**Grade boost**

In timing experiments, convert minutes and seconds into seconds (e.g. 2 min 30 s = 150 s). Data in seconds are easier to plot. Also, plotting 2 min 30 s as 230 seconds by mistake happens and ruins otherwise good answers!

Be careful to consider the following when planning graphs in practical work and examinations:

1. Axes – clearly labelled to show the quantity that is being plotted. This could be an algebraic symbol, e.g. $V$, which should be defined.

2. Scales – linear (i.e. equal intervals on the scale represent equal increments in the quantity) and chosen so that the plotted points occupy at least half of the grid in both the horizontal and vertical directions. Avoid awkward intervals – with a factor of 3 or 7. It is not always necessary to include 0.

3. Units – with the axis labels, e.g. time / s; $P$ / mW; acceleration / m s$^{-2}$.

   The following plot illustrates these points: How many mistakes can you see?

   - The vertical axis is not clearly labelled. Even if we assume that 'Temp' stands for temperature, the unit is not given.
   - The horizontal axis is not labelled, though there is a unit (seconds) so presumably it stands for time.
   - The horizontal scale is not uniform. The gap between 0 and 20 is the same as that between 20 and 30.
   - The vertical scale contains an awkward factor of 3.
   - The plotted points occupy less than half the vertical extent of the grid.

   To see the effects of these mistakes, look at point A on the grid. What values does it represent?

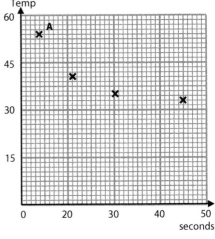

*Fig. 3.3 Poor graph planning*

4. Plotting points. Plot accurately.
   The maximum tolerance which is allowed is ± ½ a minor scale division.

5. Drawing the graph. In physics experiments, it is generally the rule that the two variables have a simple monotonic relationship (i.e. as one variable increases, the second variable either always increases or always decreases) without any sudden kinks. Because of this, a best-fit line should be drawn. Note that 'best-fit line' isn't always a straight line.

### Tip for drawing a best-fit straight line

Mark in the centroid of the data points – this is the point with the mean $x$-value, $\bar{x}$, and mean $y$-value, $\bar{y}$. If all the points are equally uncertain, the best fit line should pass through their centroid $(\bar{x}, \bar{y})$. Rotate the ruler about this point so that its edge passes as close as possible to all the points.

## 3.3.2 Extracting data from graphs

### (a) Reading off a pair of values

Drawing a best-fit line is equivalent to producing a running average of the readings. If asked to state a $y$-value for a given $x$-value, or *vice versa*, **always** take a reading from the graph, even if the given value is one of the data points – the best-fit line may not pass through the data point.

### (b) Finding the gradient and intercept

The gradient, $m$, is defined by:

$$m = \frac{\Delta y}{\Delta x}$$

On a linear graph, draw a large triangle, as shown, measure $\Delta y$ and $\Delta x$ and calculate $m$. Remember to think about the scale; here, 1 cm on the horizontal axis represents 1.0 mm.

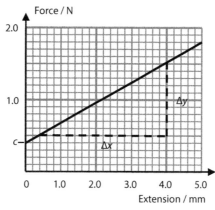

Fig. 3.4 Gradient of a linear graph

The intercept is the reading on the vertical ($y$) axis where the graph hits the axis, represented by c on the graph.

If the graph is a curve, to measure the gradient at a point, draw a tangent at that point and measure the gradient, as for a linear graph.

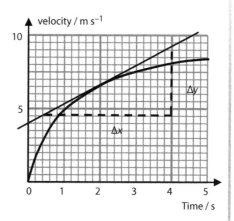

Fig. 3.5 Gradient of a non-linear graph

## Grade boost

If the graph slopes downwards, the value of $\Delta y$ is negative and so the gradient is negative.

## quickpire

④ What are the gradient and intercept of the linear graph?

## quickpire

⑤ What is the gradient of the curved graph at a time of 2.0 seconds?

# 3.4 Relationships between variables

## 3.4.1 Testing relationships using raw data

You can use data from experiments to test whether two variables are:

- proportional (or 'directly proportional')
- linearly related
- inversely proportional.

The variables are identified as $x$ and $y$ below.

(a)  Proportional: variable $y$ is (directly) proportional to $x$ – or $x$ and $y$ are (directly) proportional – if $\frac{y}{x} = k$, where $k$ is a constant. This relationship can be written:

$$y \propto x$$

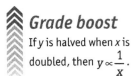

(b)  Linear relationship: if $y$ is linearly related to $x$, then equal increases in $x$ produce equal increases in $y$.

(c)  Inversely proportional: variable $y$ is inversely proportional to $x$ if $xy = k$, where $k$ is a constant. This relationship can be written:

$$y \propto \frac{1}{x}$$

(d)  Different relationships: we can use one of these three tests for other similar functions too, e.g. we can test whether $y \propto x^2$. Just use $x^2$ instead of $x$ in the first test, i.e. look at the ratio $\frac{y}{x^2}$. If it is always the same then $y \propto x^2$. Similarly, if $x^2 y$ = constant then $y \propto \dfrac{1}{x^2}$.

**Exercise**: Look at the following table of data – then do Quickfire 6.

|   | Dependent variables | | | |
|---|---|---|---|---|
| $x$ | $y_1$ | $y_2$ | $y_3$ | $y_4$ |
| 2.0 | 30.0 | 1.5 | 0.8 | 1.0 |
| 4.0 | 15.0 | 2.5 | 1.6 | 4.0 |
| 6.0 | 10.0 | 3.5 | 2.4 | 9.0 |
| 8.0 | 7.5 | 4.5 | 3.2 | 16.0 |

The problem with using raw data to test relationships is that experimental data are not perfect. This means that multiplying pairs of values of the two variables does not always give exactly the same answer, even if the two variables are inversely proportional.

## 3.4.2 Testing relationships using graphs

This is a much more powerful tool than using raw data.

The relationship between two variables is linear if a graph of $y$ against $x$ is a straight line.

The variables are related by the equation:

$$y = mx + c$$

where $c$ is the intercept on the $y$ axis and $m$ is the gradient defined by:

$$m = \frac{\Delta y}{\Delta x}.$$

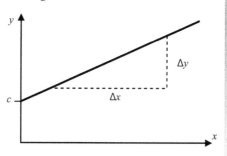

Fig. 3.6 Finding m and c

If $y \propto x$, the graph is also a straight line but the intercept, $c$, is 0.

### Example

The relationship between the terminal pd, $V$, and the current, $I$, through a power supply is:

$$V = E - Ir,$$

where $E$ is the emf and $r$ the internal resistance. If $E$ and $r$ are constant the relationship between $V$ and $I$ is linear and we can use a graph of $V$ against $I$ to determine the emf and internal resistance.

Re-writing the equation for the cell:    $V = (-r) I + E$

Comparing with the linear equation:    $y = m x + c$

The double arrows indicate that, if we plot a graph of $V$ ($y$-axis) against $I$ ($x$-axis), it should be a straight line with intercept $E$ on the $V$ axis and gradient $-r$.

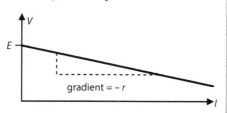

gradient $= -r$

## 3.4.3 If the relationship between the variables is not linear

Think about the equation for constant acceleration $v^2 = u^2 + 2ax$, with $v$ and $x$ the variables. This is not a linear equation because $v$ is squared, so a plot of $v$ against $x$ would not be a straight line. If we plot $v^2$ against $x$, this is what happens

Re-arranging the equation:    $v^2 = (2a) x + (u^2)$

Comparing with the linear equation:    $y = m x + c$

---

**Grade boost**

The gradient is the term that multiplies the variable on the right-hand side of the equation.

---

**Pointer**

For a constant acceleration, $v = u + at$. From a graph of $v$ against $t$, the intercept is $u$ and the gradient is $a$.

---

**quicKfire**

⑦ The diffraction grating equation can be written $d \sin\theta = n\lambda$. If $\sin\theta$ is plotted against $n$, what are the gradient and intercept? How can $\lambda$ be found?

---

**quicKfire**

⑧ (a) What is the gradient of a graph of $T^2$ against $m$?

(b) What other graph would be a straight line and how would you find $k$ from it?

## quicKfire

⑨ The relationship between the pd, $V$, across a power supply and the external resistance, $R$, can be written:

$$\frac{1}{V} = \frac{r}{ER} + \frac{1}{R}$$.

(a) What graph, with $V$ and $R$ as variables, would be a straight line?

(b) How would you use it to find $E$ and $r$?

## Pointer

The uncertainty can be expressed to 2 s.f. if the first significant figure is 1, e.g. a resistance expressed as $5.25 \pm 0.13 \ \Omega$ is acceptable.

## Grade boost

The percentage uncertainty in $\sqrt{x}$ is ½ × the percentage uncertainty in $x$. Similarly

$$p(\sqrt[3]{x}) = \frac{1}{3}p(x)$$

## Grade boost

In the sphere formula, $V = \frac{4}{3}\pi r^3$, the values of $\frac{4}{3}$ and $\pi$ are exact, i.e. they have zero uncertainty, so the percentage uncertainty in $V = 3 \times$ the percentage uncertainty in $r$.

## quicKfire

⑩ How should you express the pressure of a car tyre from the following measurements:

225 kPa, 229 kPa, 219 kPa and 213 kPa?

The graph should be a straight line with gradient $2a$ and intercept $u^2$ on the $v^2$ axis – so we can easily find $a$ and $u$ from the graph.

Another example (you will meet this at A level). The period $T$ of oscillation on a mass $m$ on a spring is given by:

$$T = 2\pi\sqrt{\frac{m}{k}}, \text{ where k is the spring constant.}$$

Now look at Quickfire 8.

# 3.5 Uncertainties

No measurements, apart from guaranteed whole numbers (e.g. the number of students in a lab) are 100% accurate. You should be aware of and, if possible, estimate, the uncertainties in all quantities you obtain through experiment. The scientific estimation of uncertainties involves probabilities but in the Eduqas A level course, the approach is simpler:

We express a result, e.g. a density determination, with its uncertainty as follows:

Density, $\rho = 2300 \pm 100 \ \text{kg m}^{-3}$. The 100 kg m$^{-3}$ is the uncertainty – strictly the 'absolute uncertainty' (see below). This means that the best estimate of the density is 2300 kg m$^{-3}$ but that its value could lie between 2200 and 2400 kg m$^{-3}$.

## 3.5.1. Estimating the absolute uncertainty in a measurement

### (a) If we only have a single measurement

The instrument resolution should be used as the uncertainty in the measurement, e.g. when measuring lengths using a metre rule, the uncertainty should be given as 0.001 m.

If we have several measurements, i.e. repeats

Suppose we make several measurements of the bounce height of a ball when dropped from 1.000 m. The readings are: 0.785 m, 0.780 m, 0.784 m, 0.787 m, 0.783 m. What bounce height do we report?

Bounce height = mean value $\pm \dfrac{\text{maximum value} - \text{minimum value}}{2}$
(i.e. mean ± ½ the spread)

$$= 0.7838 \pm \frac{0.787 - 0.780}{2} \text{ m}$$

$$= 0.7838 \pm 0.0035 \text{ m}$$

**Rule 1**: give the uncertainty estimate to 1 significant figure, i.e. 0.004 m in this case.

**Rule 2**: give the best estimate to the same number of decimal places as the estimate of the absolute uncertainty.

So, applying both these rules: Bounce height = 0.734 ± 0.004 m, which means, 'the best estimate of the bounce height is 0.734 m but it could be anywhere in a range of 0.004 m either side of this figure'.

## 3.5.2 Fractional / percentage uncertainty

When we combine quantities by multiplying and dividing, we need to consider the fractional uncertainty or precision, $p$. This is defined as:

$$p = \frac{\text{absolute uncertainty}}{\text{best estimate}}$$

This is often expressed as a percentage (say, 4% rather than 0.04).

**Example**

The emf of a cell is 1.57 ± 0.04 V. What is the fractional uncertainty?

**Answer**

$$p = \frac{0.04}{1.57} = 0.0255 = 2.55\%$$

Again this is usually expressed to 1 s.f. as 0.03 or 3% but you should keep more s.f. in the middle of calculations.

## 3.5.3 Uncertainties in derived quantities

Most equations in physics require quantities to be multiplied together or for one quantity to be divided by another.

Examples: $R = \dfrac{V}{I}$; $E_p = mgh$.

When we combine quantities in these ways, the precision in the result is the sum of the precisions in the quantities, e.g. using the first example above $p_R = p_V + p_I$ (Note: here $p_R$ means the precision in $R$, etc.)

**Example**

The percentage uncertainties in $V$ and $I$ are 2% and 3% respectively. The calculated value of the resistance is 539.2 Ω. How should this be expressed?

**Answer**

Adding the $ps$: $p_R = 2\% + 3\% = 5\%$.

The absolute uncertainty in $R$ is 5% × 539.2 Ω = 30 Ω (to 1 s.f.), so the resistance should be expressed as 540 ± 30 Ω.

Sometimes we need to raise a quantity to a power, e.g. Volume $= \frac{4}{3}\pi r^3$.

The quantity $r^3$ is just $r \times r \times r$ so, using the above rule, the percentage uncertainty in $r^3$, $p(r^3)$ is 3 × the percentage uncertainty in $r$. The general rule is $p(x^n) = n \times p_x$.

**Grade boost**

If a diameter, $d$, is measured and used to calculate the radius, $r$, then $p_r = p_d$.

**quicKfire**

⑪ An asteroid's speed is given as 35.2 ± 0.5 km s⁻¹. What is the percentage uncertainty?

**quicKfire**

⑫ The volume of a cylinder is given by $V = \pi r^2 l$.

(a) If $p_r = 0.5\%$ and $p_l = 0.8\%$, what is $p_V$?

(b) If the calculated value of $V$ is 15.34 cm³, how should the volume be expressed?

**quicKfire**

⑬ A ball bearing has a diameter of 1.00 ± 0.01 cm. What is its volume?

# 3.6 Uncertainties from graphs

You can find the uncertainty of the gradient and intercept of a linear graph by plotting the error bars rather than the mean values of data.

## 3.6.1 Uncertainty in only one variable

Consider the following experimental point: (2.50 s, 35 ± 2 m s⁻¹). The data tell us that, at a time of 2.50 s, the speed is between 33 and 37 m s⁻¹. This is plotted as on the grid: This shape is referred to as an *error bar*. The cross-bars at the ends have no function apart from showing the error bar clearly.

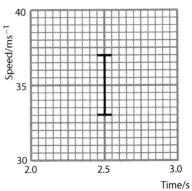

Fig. 3.7 Error bars

Some additional points:

(i) Sometimes, the examiner gives the *percentage* uncertainty rather than the absolute uncertainty, as in the table.

You need to work out the absolute uncertainty: in this case 8% of 35 is 3 m s⁻¹ (1 s.f.) so the error bar would be from 32 to 38 m s⁻¹.

| Time/s | Speed/m s⁻¹ (Uncertainty 8%) |
|--------|------------------------------|
| 2.50   | 35 ± ............            |

(ii) Sometimes the examiner will give data which need to be manipulated, e.g. finding a square, a square root or reciprocal.

| $x$/m (Uncertainty 5%) | $\left(\dfrac{1}{x/\text{m}}\right)^2$ |
|------------------------|----------------------------------------|
| 0.250                  | ............ ± ............             |

In this case $\left(\dfrac{1}{0.250}\right)^2 = 16.0$ with an uncertainty of 10%, i.e. 16 ± 2 m⁻² or 16.0 ± 1.6 m⁻².

Sometimes the data may be more complicated, e.g. the plot of $yT^2$ against $y^2$ for the compound pendulum. In this case the rules for combining the uncertainties from Section 3.5.3 hold.

**Note**

Sometimes the $x$-variable will have the uncertainty, in which case the error bar will be horizontal.

> ## Pointer
>
> Remember the following properties of uncertainties:
> - $p(xy) = p(x) + p(y)$
> - $p(x^n) = n.p(x)$
> - $p\left(\dfrac{x}{y}\right) = p(x) + p(y).$

## quickpire

(14) If $y = 0.50 ± 0.01$ m and $T = 1.88 ± 0.05$ s, between which values will the error bar for $yT^2$ be?

## 3.6.2 Uncertainty in both variables

Often there is a significant uncertainty in both variables, e.g. (2.0 ± 0.2 s, 86 ± 3 cm). In this case two error bars are plotted at right angles as shown.

As an *alternative*, some people like to plot a rectangle, or 'error box'. This is perfectly acceptable; indeed it can be very useful when drawing best-fit lines.

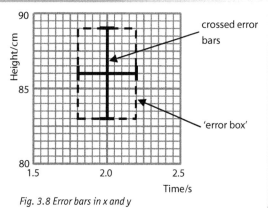

Fig. 3.8 Error bars in x and y

**Grade boost**

Before drawing the error bar(s), plot the point using one of the following:

.

×

+

This will help to ensure that the error bars are correctly positioned.

**Pointer**

The steepest and least steep lines when using error bars are often called the max/min graphs.

## 3.6.3 Drawing graphs using error bars

Once you have plotted the error bars, rather than drawing a best-fit line, you should draw the steepest and least-steep lines consistent with the data. Once you have done this you will be able to find the gradient (with its uncertainty) and the intercept (with its uncertainty).

To see how this is done, it is best to use an example. Note: In this example and the subsequent diagrams, the error bars are all shown on the $y$-variable, i.e. the error bars are all vertical. Exactly the same principles hold for horizontal error bars.

The following graph relates to the equation: $x = ut + \frac{1}{2} at^2$

If we divide by $t$ we get $\frac{x}{t} = u + \frac{1}{2} at$, so a plot of $\frac{x}{t}$ against $t$ should be a straight line. Its gradient is $\frac{1}{2} a$ and the intercept on the vertical axis is $u$.

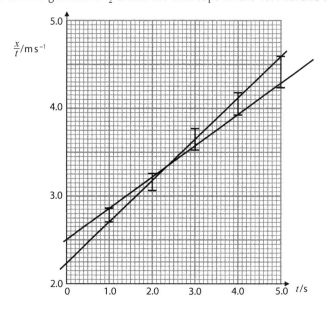

Fig. 3.9 Max/min graphs

**Grade boost**

Take care with the scales when drawing in the error bars, especially when the horizontal and vertical scales are different.

**quicKpire**

⑮ A graph of $v^2$ against $x$ from the relationship $v^2 = u^2 + 2ax$ has a gradient of 3.20 ± 0.10 and an intercept of 6.25 ± 0.15. The units m and s are used.

Give the values and uncertainties of $u$ and $a$.

## Grade boost

When finding the gradient of the steepest and least steep lines:

1. Use points on the line that are as far apart as possible.
2. Indicate clearly which points you are using, e.g. by drawing a gradient triangle or labelling them on the graph.
3. Make sure you read the scales correctly when reading the horizontal and vertical separation of the points.

## » Pointer

If the intercept on the y-axis is off the scale at the bottom or top of the grid, don't panic! You can calculate the intercept using your value of the gradient and any two (x, y) values from the line, using the equation $y = mx + c$.

**Example**
If a line has gradient 3.0 and passes through the point (50, 80), putting these values into $y = mx + c$, gives
$80 = 3.0 \times 50 + c$,
which leads to $c = -70$.

---

The sequence is:

1. Work out the scales, draw and label the axes.
2. Plot the points with error bars.
3. Draw the *steepest possible* straight line passing through the error bars (Note: Not the straight line through the tops of the error bars!)
4. Draw the least steep line through the error bars.

The aims are to *verify the relationship* and to *determine u and a*.

It is possible to draw a straight line through the error bars, so the data are consistent with a linear relationship between $\frac{x}{t}$ and $t$.

Is the relationship verified? Most scientists would say that it is not possible to show that a law or a relationship is true – but we can attempt to falsify it. In this case, we can say that the results are *consistent* with the suggested relationship because it is possible to draw a straight line passing through all the error bars.

Determining $u$ and $a$: assuming the relationship, we note that the steepest line passes through (0.00, 2.23) and (5.00, 4.59) and the least steep line passes through (0.00, 2.50) and (5.00, 4.28). Using these points, the maximum and minimum gradient values are:

$$m_{max} = \frac{4.59 - 2.23}{5.00} = 0.472 \text{ m s}^{-2} \quad \text{and} \quad m_{min} = \frac{4.28 - 2.50}{5.00} = 0.350 \text{ m s}^{-2}$$

So $$m = \frac{0.472 + 0.350}{2} \pm \frac{0.472 - 0.350}{2} \text{ m s}^{-2} = 0.41 \pm 0.06 \text{ m s}^{-2}$$

And the intercept $$c = \frac{2.50 + 2.23}{2} \pm \frac{2.50 - 2.23}{2} \text{ m s}^{-1} = 2.37 \pm 0.14 \text{ m s}^{-1}$$

Now, $u$ = the intercept. So $u = 2.37 \pm 0.14$ m s$^{-1}$ or $2.4 \pm 0.1$ m s$^{-1}$

And $a = 2 \times$ gradient $= 0.82 \pm 0.12$ m s$^{-2}$ or $0.8 \pm 0.1$ m s$^{-2}$.

## With error-bars in two directions

These are treated just the same as those in one direction. You just need to remember that the 'true' position of the point could be anywhere in the rectangle defined by the error bars, i.e. the shaded part of the diagram to the right. This is why some people use the 'error box' method of drawing the error bars – the error box is the same as the shaded area. Here is an example of a graph with $x$ and $y$ error bars.

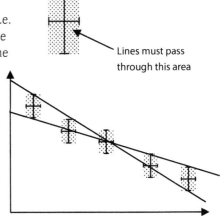

Lines must pass through this area

*Fig. 3.10 Error boxes*

## Three common mistakes with lines

1. Forcing the lines through the origin – (0,0)

   Students often do this when they expect the graph to pass through (0,0). The correct method is either to ignore zero when drawing the steepest and least-steep lines or to treat it as having error bars itself.

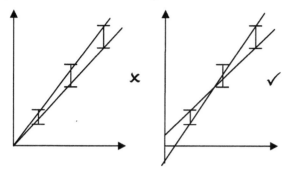

Fig. 3.11 The origin error

With these error bars, we would say that the data are consistent with a proportional relationship – it is possible to draw a straight line through the origin and all the error bars.

2. Taking the lines through the cross-bars.

   Remember that these cross-bars have no status – they just clarify the ends of the error bar.

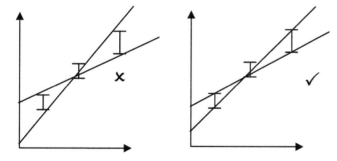

Fig. 3.12 The cross-bar error

NB – steepest and least-steep lines can go down as well as up.

3. Drawing 'tram lines'.

   This is just wrong – don't do it!

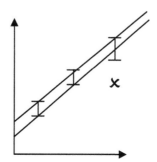

Fig. 3.13 The tram-line error

**》 Pointer**

Sometimes sneaky examiners might produce data which have a straight-line portion and a curved portion or two straight line portions. They will usually tell you to expect this:

e.g., how would you draw in the lines with these error bars?

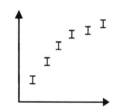

## ≫ *Pointer*

Significant areas under graphs:
- $v$–$t$ graph: displacement
- $a$–$t$ graph: change in velocity
- $F$–$x$ graph: work done / energy transfer
- $I$–$t$ graph: charge transfer
- $P$–$t$ graph: energy transfer

# 3.7 Finding the 'area under the graph'

In many physics graphs the 'area' between the graph and the $x$-axis has some physical significance. Some examples are given in the Pointer. We have put 'area' in quotes because it isn't really an area; e.g. multiplying the velocity by the time is a displacement not an area but it looks like an area on a diagram and we're going to drop the fussy quotes!

## 3.7.1 Linear graphs

Many graphs in exam papers are straight lines – often a series of straight lines. The $v$–$t$ graph in Fig. 3.14 is a typical example. There are two main approaches:

### (a) Dividing the area into triangles and rectangles

There are two obvious triangles and one rectangle in Fig. 3.14. Using the formulae:

Area of a triangle $= \frac{1}{2}\,bh$   and   area of a rectangle $= bh$

Area under the graph $= (\frac{1}{2} \times 8 \times 15) + (12 \times 15) + (\frac{1}{2} \times 6 \times 15) = 285$ m

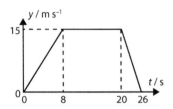

Fig. 3.14 Area under a graph

### (b) Using the area of a trapezium

A trapezium is a quadrilateral with a pair of parallel sides. Its area is the mean length of the parallel sides multiplied by their separation.

Applying this to Fig. 3.15, we have $a = 12$ s, $b = 26$ s and $h = 15$ m s$^{-1}$.

∴  Area $= \frac{1}{2}\,(12 + 26) \times 15 = \frac{1}{2} \times 38 \times 15 = 19 \times 15 = 285$ m

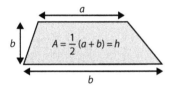

Fig. 3.15 Area of a trapezium (1)

Note that the parallel sides of a trapezium do not need to be horizontal. We shall use vertical-sided trapeziums in Section 3.7.2(c). This should remind you of the derivation of $x = \frac{1}{2}\,(u + v)t$ , for a very good reason!

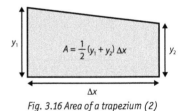

Fig. 3.16 Area of a trapezium (2)

## 3.7.2 Non-linear graphs

The curve in Fig. 3.17 is a typical non-linear graph. It is the sort of graph you'd expect when plotting a load-extension curve for a rubber band. The area under the graph represents the work done in extending the rubber band. There are several ways of estimating this area.

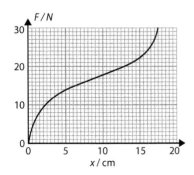

Fig. 3.17 Non-linear graph

## (a) Drawing an estimated straight line

The dotted line in Fig. 3.18 is an attempt by the author to produce two equal areas, indicated by *, one of which is below the curve and the other above. The area under the pecked line should then be the same as that below the curve.

In this case the area, A, under the pecked line is given by

$A = \frac{1}{2}bh = \frac{1}{2} \times 0.16$ m $\times 30$ N $= 2.4$ J

The author's guess does not turn out to be particularly accurate (see below). Perhaps you could do better.

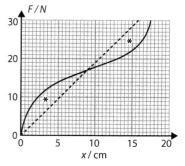

Fig. 3.18 Estimated straight line

## (b) Counting squares

This is quite a quick method. The squares in Fig. 3.19 have bases of 0.025 m and heights of 5.0 N, so the area of each is:

$$0.025 \text{ m} \times 5.0 \text{ N} = 0.125 \text{ J}.$$

We count the squares below the line. But what do we do about part squares? More than half counts as a whole; less than half counts as 0. This, again, requires some judgement.

In the author's judgement there are 22 such squares.

∴ Estimate of area $= 22 \times 0.125$ J $= 2.75$ J

This can be made more accurate by choosing the smaller (0.005 m by 1.0 N) squares – or a combination of the two.

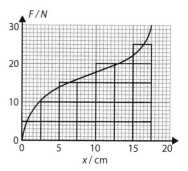

Fig. 3.19 Square counting

## (c) Dividing into trapeziums[1] – the 'trapezoidal rule'

This is potentially quite accurate but takes a bit longer than the other two methods. The graph is approximated by a series of straight lines – not just one as in method (a). The area of each is calculated using the formula for the area of a trapezium (in this case, the mean of the two heights times the base) and the areas added together.

Fig. 3.20 shows this done by dividing the graph into equal intervals $\Delta x$.

If this is done and the $y$ values are called $y_1, y_2 \ldots y_N$, the total area, A, is given by:

$$A = \frac{1}{2}(y_0 + 2y_1 + 2y_2 + \ldots + 2y_{N-1} + y_N)\,\Delta x$$

### Example

Use the trapezoidal rule to estimate the work done in Fig. 3.17 in stretching the rubber band by 17.5 cm

### Answer

Using the trapeziums in Fig. 3.20:

Work done $\quad = \frac{1}{2}(0 + 2(10.5 + 13.5 + 16 + 17.5 + 20 + 22) + 30)\,0.025$

$\quad\quad\quad\quad = 2.86$ J

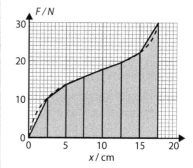

Fig. 3.20 Using trapeziums

---

[1]  We decided against 'trapezia'!

# Exam practice and technique

## Answering exam questions

To some students, a question is a question[1] is a question. It is there to be answered and they don't worry themselves about how it is put together and what skills they'll need to demonstrate in answering it. However, you'll expect to gain more marks if you are able to read a question and understand the sort of answer the examiner is expecting (or at least, hoping for).

When setting questions, examiners work within certain rules so that an exam paper one year tests the same sort of abilities as in other years – without using the same questions! The main constraints on questions are the 'Assessment Objectives' (AOs).

### Assessment objectives

**AO1**: 35% of the marks are showing that you recall and understand aspects of physics, e.g. you can state a law or definition, you know what equation to use to solve a problem or you can describe how you would carry out an experiment. (In the full A Level it is 30% of the marks.)

**AO2**: 45% of the marks are for using the AO1 knowledge in order to solve problems. This involves producing answers to calculations, bringing together ideas to explain things, combining and manipulating formulae, using experimental results and graphs.

**AO3**: 20% of the marks are for such things as reaching conclusions from experimental results or other data and developing experimental techniques. (In the full A Level it is 25% of the marks.)

### Skills

At the same time as balancing the AOs, the examiner looks at the balance of skills. Being physics, a high percentage of marks (at least 40%) comes from the application of **mathematics**.

This might actually seem a little low but the low-level skill of putting numbers into a formula doesn't count as a mathematical skill but is essentially just communication!

On the other hand, drawing tangents and finding their gradients is certainly maths. In any case, examiners are free to demand more than 40% maths skills, without being excessive.

**Practical skills** account for at least 15% of the assessments. Such questions involve designing, analysing and drawing conclusions from **experiments**, e.g. from the specified practical work.

As an illustration, look at this multi-part question.

(a) A student uses a metre rule, a pivot, a 0.2 N weight and a test tube to measure the mass of a ball bearing. He sets up the apparatus as in the diagram:

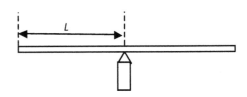

---

[1] Actually, most exam 'questions' are not questions! They are instructions to do something: *State Newton's second law of motion; describe how a laser works.* But we'll call them questions anyway.

(i) He adjusts the ruler until it is balanced at the centre of gravity. State what is meant by 'centre of gravity'. [1]

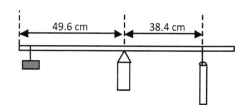

(ii) He finds the length, $L$, to be 49.6 cm. He hangs the 0.2 N weight 4.0 cm from the end of the rule and then puts on the empty test tube, adjusting its position until the apparatus balances, as shown.

With the ball bearing in the test tube, the student finds the balance length between the test tube and pivot found to be 24.5 cm.

Show in clear steps that the mass of the ball bearing is approximately 14 g. [4]

(b) The student measures the diameter of the ball bearing to be 1.50 cm. A science data book gives the following data for the densities, in kg m$^{-3}$, of some metals:

Aluminium, 2800    Iron 7950    Copper 8900    Lead 11 300

Determine the material of the ball bearing and justify your answer. [4]

Within this 9-mark question there are: 3 AO1 marks, 2 AO2 marks, 4 AO3 marks, 6 maths marks and 8 practical marks. Can you spot how these marks are allocated? The answer is given at the end of the Quickfire answers.

## Style

Question papers are a mixture of several kinds of questions including

- Short answer, e.g. a single sentence
- Extended response, i.e. a paragraph
- Calculations
- Graph drawing

# Exam tips

Now we'll have a look at some tips to enable you to show what you do in answering exam questions. The first and most important point is to read the question carefully. Examiners discuss the wording of questions so that the meaning is clear and precise. In spite of this, it is easy to misinterpret a question so take your time. Using a highlighter to mark key information often helps, e.g. numerical information given at the beginning of a question is sometimes not needed until later on, and so highlighting this makes it stand out.

## Look at the mark allocation

Each part of a question is allocated a number of marks. In written answers, this total gives a hint as to how much detail you need in your answer. In calculations, some marks will be for the working and some for the answer [see below].

## Understand the command words

These are the words which show the sort of answer which the examiner expects in order to give you credit.

### State

A short answer with no explanation.

## Explain

Give a reason or reasons. Look at the mark allocation: 2 marks usually means that you need to make two distinct points.

## State and explain

There may be a mark for the statement but the first mark may be for an explanation of a correct statement, e.g. *'State which resistor, A or B, has the higher value and explain your reasoning.'* It is unlikely that the examiner will give you a mark for a 50/50 choice!

## Calculate

A correct answer will score all the marks, unless the question includes the instruction to *'show your working.'*
**Warning**: An incorrect answer without working will score 0.

## Show that [in a calculation question]

E.g. 'Show that the resistivity is approximately $2 \times 10^{-7}$ $\Omega$ m.' There is no mark here for the correct answer; the working must be shown in sufficient detail for the examiner to be convinced you know what you are doing!
**Hint**: In this case, calculate an accurate answer, e.g. $1.85 \times 10^{-7}$ $\Omega$ m and say that this is approximately the value stated.

## Describe

A series of statements is required. These may be independently marked but care may be needed with sequencing, e.g. in the description of how to carry out an experiment.

## Compare

There must be a clear comparison, not just two separate statements. It is also not safe just to state one thing and leave the examiner to infer another; e.g. *'Compare the work functions of metals A and B.'*

Answer 1: Metal A has a low work function – not enough.

Answer 2: Metal A has a lower work function **than metal B** – this answer would gain credit (if correct!) unless the question makes it clear that a numerical comparison is required.

## Suggest

This command word often comes at the end of a question. You are expected to put forward a sensible idea based upon your physics knowledge and the information in the question. There will often not be a single correct answer.

## Name

A single word or phrase is expected; e.g. 'Name the property of light being demonstrated' (*in a question showing waves spreading out after passing through a gap*). Answer: *Diffraction*. Note that, especially in this kind of question, a correctly spelled answer may be required.

## Estimate

This does not mean 'guess'. It usually involves one or more calculations with simplifying assumptions. The question may ask you to state any assumptions you make. E.g. *Estimate the number of 1 mm diameter spheres which will fill a measuring cylinder up to the 100 cm³ mark.*

## Derive

This involves producing this equation starting from a set of assumptions and

or more basic equations. These are: $v = u + at$, $x = ut + \frac{1}{2}at^2$, $I = nAve$, $E_{elastic} = \frac{1}{2}kx^2$ and $\dfrac{V_{OUT}}{V_{IN}} = \dfrac{R}{R_{total}}$

You should learn the derivations of these equations.

**Discuss**

This command word is often used in the context of social and ethical issues. Ideally, an answer should contain at least two contrasting points of view together with supporting statements. There is no requirement for the discussion to be balanced.

## Tips about diagrams

Questions about experiments sometimes ask for diagrams. The diagram should show the arrangement of the apparatus and be labelled. Separate diagrams of a metre rule, a length of wire, a micrometer and an ohm-meter will gain no credit. Note, however, that standard circuit symbols, e.g. a cell or a voltmeter, do not need labelling. Even if the question does not demand one, some of the marks may be awarded for information included in a well-drawn diagram.

## Tips about graphs

**Graphs from data**: Where the axes and scales are not drawn, make sure that the scale occupies most or all of the given grid and that the plotted points occupy at least half of the height and width of the grid. Label the axes with the name, or symbol, of the variable with its unit – e.g. time / s, or $F$ / N – and include scales. Plot points as accurately as possible; for points requiring interpolation between grid lines, the usual tolerance is ± ½ a square. Unless the question instructs differently, draw in the graph, don't just plot the points.

**Sketch graphs**: A sketch graph gives a good idea of the relationship between the two variables. It needs labelled axes but often it will not have scales and units. It is **not** an untidy ('hairy') graph. If the graph is intended to be a straight line, it should be drawn using a ruler. Sometimes significant values need to be labelled, e.g. include $f_S$ and $\phi$ on a sketch graph of photoelectron energy against frequency.

## Tips about calculations

If the command word is **calculate**, **find** or **determine**, full marks are given for the correct answer with no working shown. But an incorrect answer with no working scores 0. There are usually marks available for correct steps in the working even if the final answer is incorrect. Points the examiner will look for include:

- Selection of equation or equations and writing them down.
- Conversion of units, e.g. hours into seconds, mA into A.
- Insertion of values into equation(s) and manipulation of equation.
- Stating the answer.
- **Remember the unit:** missing or incorrect units will be penalised.

## Tips about describing experiments

When describing one of the experiments from the specified practical work or for a practical which you are devising as part of the examination:

- Draw a simple diagram of the apparatus used **in its experimental arrangement**.
- Give a clear list of steps.
- Say what measurements are made and which instrument will be used.
- Say how the final determination will be made from your measurements.
- If required, give precautions or techniques to give accurate results.

## Tips about questions involving units

This is mainly dealt with in Section 1.1.1 – Units and dimensions. One type of question requires you to suggest a unit for a quantity, which might not be on the specification. Such questions will always give an equation involving the quantity. The procedure to adopt is:

- Manipulate the equation to make the unknown quantity the subject.
- Insert the known units for the other quantities.
- Simplify.

### Example

The drag, $F$, on a sphere of radius $a$ moving slowly with a velocity, $v$, through a fluid is given by $F = 6\pi\eta av$ where $\eta$ is a constant called the coefficient of viscosity. Suggest a unit for $\eta$.

### Answer

Make $\eta$ the subject: $\qquad \eta = \dfrac{F}{6\pi av}$

Rewrite in terms of units: $\therefore$ unit of $\eta = \dfrac{N}{m \times m\,s^{-1}} = N\,m^{-2}\,s.$

**Note**: The question did not ask for any particular form, e.g. reducing it to the base units, so it can be left like this. The pascal, Pa, is equivalent to $N\,m^{-2}$ so another equivalent is Pa s.

## QER questions

Every examination paper will contain a question, or part of a question, which will test your ability to present a coherent account. These are called **Quality of Extended Response** questions and they are worth 6 marks. They could be AO1 questions on a piece of bookwork, e.g.

> Explain how a laser works $\qquad\qquad\qquad$ [6 QER]

or a description of one of the pieces of specified practical work, e.g.

> Explain in detail how you would determine the Young modulus of a metal in the form a long wire.
> $\qquad\qquad\qquad$ [6 QER]

It could also ask you to apply you knowledge to a specific situation, so it includes some AO2, e.g. (following a diagram of energy levels in a laser system):

> Explain in detail how light amplification takes place for the above laser system.
> $\qquad\qquad\qquad$ [6 QER]

Whatever the topic of the question, the examiner will be looking for an identified set of ideas connected into 'a sustained line of reasoning which is coherent, relevant, substantiated and logically structured'. This means that the inclusion of incorrect or irrelevant material or poorly constructed arguments will be penalised. See the Exam Q&A section for an example of good and less good answers.

## Synoptic assessment

Every question paper will include some marks for concepts which are specified in other components. The whole question will not focus on these concepts but may apply them to the content of the component which is being examined. As an example, questions in any component could explore the homogeneity of equations, which is specified in component 1.

# Practice questions

## Definition-type questions

1. A body is acted on by a number of forces. State the conditions needed for the body to be in equilibrium
   
   *[Basic physics – 1.1]*

2. In terms of electromagnetic radiation, a star can be considered as a *black body*.
   What is meant by the term 'black body'?
   
   *[Using radiation to investigate stars – 1.6]*

3. Progressive waves can be either *transverse* or *longitudinal*.
   Distinguish between transverse and longitudinal waves and give an example of each.
   
   *[The nature of waves – 2.4]*

4. 'The potential difference across a component is 5.0 V.'
   Explain this statement in terms of energy.
   
   *[Resistance – 2.2]*

5. State Ohm's Law.
   
   *[Resistance – 2.2]*

6. State the meaning of the term 'transition temperature' when applied to electrical conductors.
   
   *[Resistance – 2.2]*

7. State the Principle of Superposition.
   
   *[Wave properties – 2.5]*

8. *Hadrons* are composite particles, which are classified as either *baryons* or *mesons*.
   State what is meant by each of the terms in italics and give an example of a baryon and a meson.
   
   *[Particles and nuclear structure – 1.7]*

9. If data are transmitted over long distances through multimode fibres, different pieces of data can be received at the same time.
   State the name for the cause of this problem.
   
   *[Refraction of light – 2.6]*

10. An electrical power supply has an *EMF of 10.0 V* and the *pd across its terminals is 9.0 V*.
    Explain the expressions in italics in terms of energy.
    
    *[DC circuits – 2.3]*

## Experimental description questions

11. (a) Starting from a defining equation, show that the unit of resistivity is $\Omega$ m.
    (b) Describe an experiment to determine the resistivity of a metal in the form of a metal wire.
       *[Resistance – 2.2]*

12. Describe an experiment to investigate how the resistance of a metal wire varies with temperature. Sketch a graph to show the expected variation.
    *[Resistance – 2.2]*

13. The diagram shows a circuit containing a vacuum photocell. The metal surface in the photocell is irradiated with electromagnetic radiation of fixed frequency.

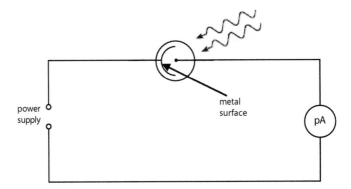

(a) Describe how you can determine the maximum kinetic energy of the photoelectrons. You should add any necessary labels and components to the diagram.

(b) In a further experiment, the maximum kinetic energy of the emitted electrons is measured for a range of frequencies of incident radiation. A graph of the results is as follows:

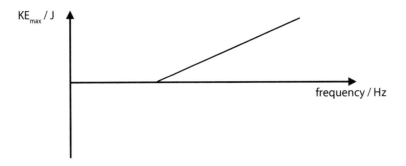

From the graph, state how you would determine:
    (i) the Planck constant;
    (ii) the work function of the metal.
    *[Photons – 2.7]*

## Questions to test understanding

14. It is suggested that the speed, $v$, of transverse waves on a stretched string is related to the tension, $T$, and the mass per unit length, $\mu$, of the string, by the equation

$$v = k\sqrt{\frac{T}{\mu}},$$

where $k$ is a dimensionless constant. Show that this is possible.

    [Basic Physics – 1.1]

15. The graph shown is of a skydiver who reaches terminal velocity then opens the parachute.

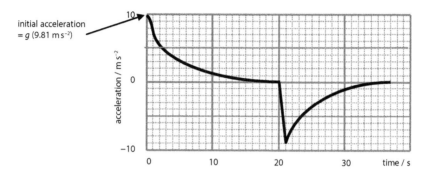

    (a) Explain the form of the graph between 19 s and 25 s.
    (b) Explain why the area between the line and the time axis must be overall positive.
    [Kinematics – 1.2]

16. Three resistors are connected in a network as shown.

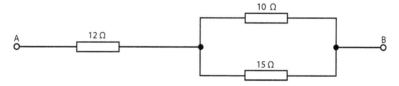

    A power supply of EMF 9.0 V and negligible internal resistance is connected between A and B.
    Calculate the pd across the 12 Ω resistor.
    [DC circuits – 2.3]

17. A cell has EMF 1.5 V and internal resistance 0.5 Ω. It is connected across a 12.5 Ω resistor.
    (a) Calculate the current in the circuit.
    (b) A number of identical such cells are to be connected in series across the same resistor in order that the current is at least 0.60 A.
    Calculate the minimum number of cells required.
    [DC circuits – 2.3]

18. A skydiver falling through the air experiences a drag force which is directly proportional to the square of her velocity.

$$F = k\,v^2.$$

(a) Show that a suitable unit for $k$ is $N\,m^{-2}\,s^2$ and express this unit in terms of the base SI units, m, kg and s, only.

(b) A skydiver has a mass of 75 kg. When she is falling at 20 m s$^{-1}$ her acceleration is 8.2 m s$^{-2}$.

    (i) By considering the resultant force on the skydiver at 20 m s$^{-1}$, show that her terminal velocity is approximately 50 m s$^{-1}$.

    (ii) Estimate the time it takes her to accelerate from 35 m s$^{-1}$ to 45 m s$^{-1}$. Explain your working clearly.

*[Kinematics – 1.2]*

19. The diagram shows a potential divider used as a power supply.

Show that $\dfrac{V_{OUT}}{V_{IN}} = \dfrac{R}{R+X}$ , stating your assumption:

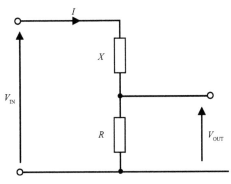

*[DC circuits – 2.3]*

20. In the following circuit the lamps, $L_1$, $L_2$ and $L_3$, are identical. The switch, S, is closed and all lamps are on. You may assume that the switch has negligible resistance.

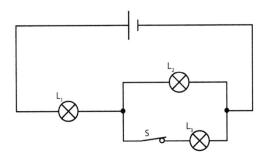

(a) Compare the brightness of the 3 lamps, $L_1$, $L_2$ and $L_3$.

    Explain your answer.

(b) Switch S is now opened so lamp $L_3$ is off. State what effect this has on the brightness of $L_1$ and $L_2$ and explain your answer.

*[DC circuits – 2.3]*

## Data analysis questions

21. A student was given a ream of A4 paper [500 sheets] and used a 30 cm rule with a 1 mm scale to determine the volume of a single sheet of A4 paper.

    She was able to use a hand lens to estimate the length and width measurements to fractions of a mm but this was not possible for the thickness measurement.

    Her measurements are:      Thickness of 500 sheets tightly held = 5.15 cm

    Length measurements = 29.72, 29.71, 29.71, 29.70, 29.71 cm

    Width measurements = 21.03, 21.05, 21.05, 21.04, 21.04 cm

    (a) Use the data to explain why the uncertainties in the length and width need not be considered when calculating the volume.

    (b) Use the student's data to calculate a value for the volume of a single sheet together with its absolute uncertainty.

22. A student is to determine the emf and internal resistance of a cell and proposes to use the circuit shown.

    Theory predicts that the current, $I$, is related to the external resistance, $R$, by the equation.

    $$\frac{1}{I} = \frac{R}{E} + \frac{r}{E}$$

    where $E$ is the emf of the cell and $r$ the internal resistance.

    The student used three 3.9 Ω resistors which he combined in various ways to act as $R$ in the circuit.

    The student's results are in the table.

    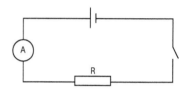

    (a) Draw diagrams to show how the student could obtain the following resistances from combinations of 3.9 Ω resistors:

        (i)   1.95 Ω

        (ii)  2.6 Ω

    (b) Complete the table.

    (c) Plot a graph of $1/I$ on the vertical axis against $R$ on the horizontal axis.

    (d) Comment on how well the graph supports the theoretical equation.

    (e) By taking suitable measurements on the graph, determine the emf and internal resistance of the cell.

| $R/\Omega$ | $I/A$ | $\dfrac{I}{I/A}$ |
|---|---|---|
| 1.3 | 0.940 | |
| 1.95 | 0.667 | |
| 2.6 | 0.533 | |
| 3.9 | 0.373 | |
| 5.85 | 0.250 | |

**Note: It is unlikely that this question would appear in its entirety on an AS exam paper but every individual aspect could.**

## Quality of extended response (QER) question

**The marking of these questions is not on the basis of individual marking points but on the quality of the overall answer.**

**In this question, part (a) is traditional and part (b) a QER question.**

23. (a) A simplified energy level diagram is shown for a 4-level laser system. The arrows show the sequence of transitions which electrons make between leaving the ground state and returning to it.

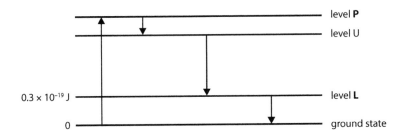

   (i)  Label the transitions associated with (I) *pumping* and (II) *stimulated emission*.

   (ii) The wavelength of the output radiation from the laser is $1.05 \times 10^{-6}$ m.

   Calculate the energy **above the ground state** of level U.

   (b) Explain in detail how light amplification takes place for the above laser system.                    [6QER]

# Questions and answers

This part of the guide looks at actual student answers to questions. There is a selection of questions covering a wide variety of topics. In each case there are two answers given; one from a student (Seren) who achieved a high grade and one from a student who achieved a lower grade (Tom)[1]. We suggest that you compare the answers of the two candidates carefully; make sure you understand why one answer is better than the other. In this way you will improve your approach to answering questions. Examination scripts are graded on the performance of the candidate across the whole paper and not on individual questions; examiners see many examples of good answers in otherwise low-scoring scripts. The moral of this is that good examination technique can boost the grades of candidates at all levels.

## Component 1: Motion, Energy and Matter

## Component 2: Electricity and Light

---

[1] The answers to question 11, the QER question, are based on student answers to this trial-run question.

**Q & A 1**

(a) (i) Define work. [2]

(ii) Hence express the unit of work, J, in terms of the SI base units kg, m and s. [2]

(b)

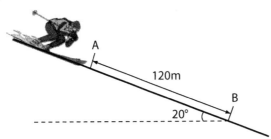

A 120m B 20°

A skier of mass 70 kg descends a slope as shown. The skier passes point **A** at a speed of 6 m s⁻¹ and **B** at 21 m s⁻¹. Calculate for the descent from **A** to **B**:

(i) The gravitational potential energy lost by the skier. [2]

(ii) The kinetic energy gained by the skier. [3]

(c) (i) State the principle of conservation of energy. [1]

(ii) Discuss your answer to (b) (i) and (ii) in terms of this principle. [2]

(d) Calculate the mean resistive force experienced by the skier between **A** and **B**. [4]

## Seren's answer

(a) (i) The force exerted multiplied by the distance✓ travelled in the direction of the force. ✓

(ii) $[F] = N = kg\ m\ s^{-2}$

$[d] = m$ ✓

$\therefore [W] = J = [F] \times [d] = kg\ m^2\ s^{-2}$ ✓

(b) (i) $\Delta gpe = mg\Delta h$

$= 70 \times 9.81 \times (120 \sin 20)$ ✓

$= 28183.82739\ J = 28.2\ kJ\ (3\ s.f.)$ ✓

(ii) $\Delta KE = \left(\frac{1}{2}mv_2^2\right) - \left(\frac{1}{2}mv_1^2\right)$ ✓

$= \left(\frac{1}{2}70 \times 21^2\right) - \left(\frac{1}{2}70 \times 6^2\right)$ ✓

$= 14175\ J = 14.2\ kJ$ ✓

(c) (i) Energy can neither be created nor destroyed, but converted into other forms of energy. ✓

(ii) The gravitational potential energy that has been lost between A and B has been converted into other forms of energy, ✓ including thermal, kinetic and sound.

(d) $Fx = \frac{1}{2}mv^2 - \frac{1}{2}mu^2$ ✓

$120\ F = \frac{1}{2}70 \times 21^2 - \frac{1}{2}70 \times 6^2$ ✗

$120\ F = 14175$ ✓

$F = 118\ N\ (3\ s.f.)$ ✗

## Examiner commentary

(a) (i) Seren has correctly included the often-forgotten statement about direction.

(ii) The first mark is for correctly identifying the force and distance units and the second for the manipulation. Note: Seren has used square brackets, [. . .], to indicate 'the unit of', which is quick.

(b) (i) Seren has gained 1 mark for correct substitution and 1 mark for the answer. In this case, converting to kJ and expressing to 3 s.f. gained no extra credit.

(ii) Fully correct. Note that, if Seren had slipped up in the final calculation, she would have received 2 marks because her working was clear and correct.

(c) (i) A mark which most candidates achieve.

(ii) Seren has identified that the initial GPE was converted into several other forms [i.e. not just kinetic] but has missed out any discussion of mechanism, e.g. work done against friction/drag.

(d) Seren has used the principle: Work done = energy transfer and used it to calculate a force from the change of KE. Unfortunately, she has calculate the resultant force on the skier, not the resistive force.

**Seren gains 13 out of 16 marks.**

## Tom's answer

(a) (i) The amount of energy used over a certain time ✗

  (ii) $J = \dfrac{kg \times m}{s}$ ✗

(b) (i) $\Delta E_p$ $= mg\Delta h\ 120 \cos 20 = 112.76\ m$
      $= 70 \times 9.81 \times 112.76$ ✓ e.c.f.
      $= 77432\ J$ ✗

  (ii) $KE = \dfrac{1}{2}mv^2$

      $= \dfrac{1}{2}70 \times 15^2 = 35 \times 15^2$ ✓ $= 7875\ J$ ✗

(c) (i) Energy cannot be created or destroyed, only transferred into other forms. ✓

  (ii) The gravitational potential energy gained by the skier has been converted into kinetic energy. ✗

(d) $a = \dfrac{v^2 - u^2}{240} = \dfrac{21^2 - 36}{240} = 1.6875\ m\ s^{-2}$

   $F = ma = -70 \times 1.6875 =$ ✓

## Examiner commentary

(a) (i) Tom's definition is of power rather than work.

  (ii) Tom's answer follows on from his confusion between work and power. *Error carried forward* would not be applied here, even if his working was correct for power, as it is a mistake of principle.

(b) (i) Tom has quoted the correct equation for $\Delta E_p$ and the first mark is for correctly using it. Unfortunately he has used cos 20° which gives the horizontal rather than the vertical distance so he just gets one mark for substitution.

  (ii) Tom has quoted the KE formula and substituted a speed so receives 1 mark. Unfortunately he has used the speed difference for $v$ rather than working out the two values of $\frac{1}{2}mv^2$ and finding the difference.

(c) (ii) Tom has not done enough for the first mark. He needed to account for the fact that gain in kinetic energy was less than the loss in gravitational potential energy.

(d) Tom has calculated the acceleration and, apart from the minus sign, has almost calculated the <u>resultant</u> force on the skier. This is the start of a possible method and has attracted 1 mark.

**Tom gains 4 out of 16 marks.**

(a) A velocity–time graph is given for a body which is accelerating in a straight line.

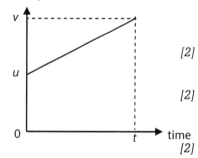

   (i) Using the symbols given on the graph, write down an expression for the gradient and state what it represents. [2]
   (ii) Using the symbols given on the graph, write down an expression for the area under the graph and state what it represents. [2]
   (iii) Hence or otherwise show clearly that, using the usual symbols,
   $$x = ut + \frac{1}{2}at^2$$ [2]

(b) A cyclist accelerates **from rest** with a constant acceleration of 0.50 m s⁻² for 12.0 s. Calculate:
   (i) the distance travelled in this time [2]
   (ii) the maximum velocity attained. [2]

(c) After 12.0 s, the cyclist stops pedalling and 'freewheels' to a standstill with constant deceleration over a distance of 120 m.
   (i) Calculate the time taken for the cyclist to decelerate to a stand-still. [2]
   (ii) Calculate the magnitude of the cyclist's deceleration. [2]

(d) Draw an acceleration–time graph on the grid for the **whole of the cyclist's journey**. [4]

(e) In reality the cyclist would not slow down with constant deceleration. This is because the total resistive force acting on the cyclist consists of a constant frictional force of 8.0 N **and** an air resistance force which is proportional to the square of the cyclist's velocity.
   (i) When the cyclist was travelling with maximum velocity, the total resistive force acting was 165 N. Calculate the force of air resistance at this velocity. [1]
   (ii) Hence calculate the total resistive force acting when the cyclist is moving at half the maximum velocity. [2]

## *Seren's answer*

(a) (i) $\frac{v-u}{t} = a$, acceleration ✓✓

(ii) $ut + \frac{1}{2}(v-u)t$ ✓ = distance travelled ✓

(iii) Distance travelled, $x = ut + \frac{1}{2}(v-u)t$

$\frac{v-u}{t} = a$ so $v - u = at$ ✓

Substituting for $(v - u)$:

$x = ut + \frac{1}{2}att = ut + \frac{1}{2}at^2$ ✓

(b) (i) $ut + \frac{1}{2}at^2$

$0 + \frac{1}{2} \times 0.5 \times 144 = 36$ m ✓✓

(ii) $0.5 \times 12$ s $= 6$ m/s ✓✓

(c) (i) $s = \frac{1}{2}(u+v)t$

$t = \frac{120}{3} = 40$ s ✓✓

(ii) $\frac{6}{40} = 0.15$ m s$^{-2}$. ✓✓

(d) acceleration (m/s$^2$)

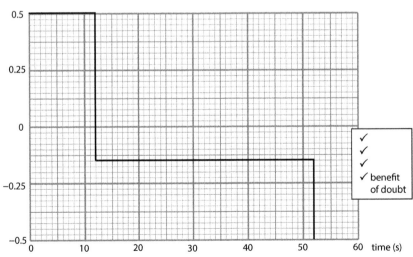

(e) (i) 157 N ✓

(ii) $157 \times 1^2 \rightarrow 157 \times 0.5^2 = \frac{157}{4} = 39.25$ N $+ 8$ N $= 47.25$ N ✓✓ [In spite of the setting out!]

## *Examiner commentary*

(a) (i) Ideally Seren would have started her answer, 'Gradient =. . . .' but that is assumed by the marker and full marks awarded.

(ii) Seren gives a fully correct answer. She doesn't simplify the expression to $\frac{1}{2}(v+u)t$, but that is not a requirement of the question.

(iii) Seren's working is made easier by the fact that she didn't simplify the expression for $x$ in part (ii)! This is a 'show that' question and Seren has explained her steps clearly.

(b) (i) As in (a)(i), the '$x =$' is missing but the working is clear, concise and correct.

(ii) Unlike Tom, Seren has used the easy method here: $v = u + at$. She doesn't write the symbol equation but this doesn't matter, as the command word is 'calculate,' so a correct answer, even with no working, would be accepted. This is also the case in (c) (ii)

(d) Scales ✓, horizontal lines ✓✓, change at 12 s. ✓ Seren has incorrectly included a vertical line down to –0.5 m s$^{-2}$ at 52 seconds but the examiner has decided to give her 'benefit of doubt' as it wasn't mentioned in the mark scheme.

(e) (i) Seren correctly subtracts 8 N to calculate the air resistance at the maximum velocity.

(ii) Seren has arrived at the correct answer and her reasoning is clear. The mathematical nonsense of $\frac{157}{4} = 39.25$ N $+ 8$ N is tolerated. Better would have been:

Air resistance force at half the maximum velocity

$= \frac{157}{4} = 39.25$ N

∴ Total resistive force $= 39.25$ N $+ 8$ N $= 47.25$ N

**Seren gains 21 out of 21 marks.**

## Tom's answer

(a) (i) The gradient represents acceleration. ✓

$(v - u) \times t = \text{gradient}$ ✗

(ii) Area under graph is the displacement ✓

$$x = \frac{1}{2}(u + v)t \checkmark$$

(iii) $v = u + at$ $\quad x = \frac{1}{2}(u + v)t$ ✓

$$x = \frac{1}{2}ut + vt + u + at \quad \text{✗ Incorrect}$$
$$\text{working}$$

$$x = ut + \frac{1}{2}at^2$$

(b) (i) $s = ut + \frac{1}{2}at^2$ ✓

$$s = 12 + \frac{1}{2} \times 0.5 \times 12^2 \text{ ✗}$$

$$s = 48 \text{ m}$$

(ii) $v^2 = u^2 + 2as$ ✓

$$v^2 = 0^2 + 2 \times 0.5 \times 48$$

$$v^2 = 48 \quad v = \sqrt{48} = 6.93 \text{ m s}^{-1} \checkmark \text{ e.c.f.}$$

(c) (i) $v^2 = u^2 + 2as$

$$6.39^2 = 0^2 + 2 \times ? \times 120$$

$$48 = 2 \times ? \times 120$$

$$\frac{48}{2 \times 120} = 0.1 \text{ m s}^{-2} \text{ ✗}$$

(ii) $0.1 \times 9.81 = 1.59 \text{ N}$ ✗

(d) acceleration (m s$^{-2}$)

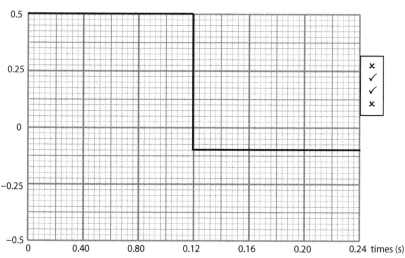

(e) (i) $165 - 8.0 = 157 \text{ N}$ ✓

(ii) $6.9 = 165 \text{ N}$

$$3.45 = 82.5 \text{ N} \text{ ✗}$$

## Examiner commentary

(a) (i) Tom successfully identified the meaning of the gradient but in his attempt at using the symbols he multiplied by $t$ rather than divided.

(ii) Tom's answer was totally correct.

(iii) The two equations Tom has written will, if correctly combined and $v$ eliminated, lead to the equation $\boldsymbol{x} = ut + \frac{1}{2}at^2$. Hence he gains the first mark. His algebra lets him down, however.

(b) (i) The correct equation is selected – the use of $s$ rather than $\boldsymbol{x}$ is accepted. Tom makes the common mistake of writing $0 \times 12 = 12$ and hence finds the displacement to be 48 m rather than 36 m.

(ii) Tom's working here is correct. It is not the easiest method of calculating $v$ [$v = u + at$ is more straightforward], but the equation he uses is correct. He has used his [incorrect] value of $\boldsymbol{s}$ but is credited with full marks on the e.c.f. principle.

(c) Tom appears to be calculating the acceleration in part (i). If he had gone on to use this to calculate $t$ he could have received credit for both parts. However, he doesn't and in part 2 goes on to calculate a force instead of the acceleration and unfortunately receives no credit at all.

(d) Tom receives marks for the two horizontal portions of the graph. He receives no credit for the scales because the scale on the time axis is not valid – and the graph doesn't encompass the whole journey.

(e) (i) Tom correctly calculates the value of air resistance at the maximum velocity.

(ii) Tom has just divided the maximum air resistance by 2. The air resistance should have been divided by 4 and the 8 N frictional force added to it.

**Tom gains 10 out of 21 marks.**

Q&A 3

A heavy sledge is pulled across a level snowfield by a force $F$ as shown. To keep the sledge moving at a constant velocity a **horizontal** force component of 200 N is required.

40°    - - - -► 200 N

(a) Calculate the force needed to produce a horizontal component of 200 N on the sledge. [2]

(b) (i) Define work done and use this definition to explain why no work is done in the vertical direction. [3]

   (ii) It takes 30 minutes to pull the sledge a distance of 2.0 km across level ground. Calculate:
      (I) the work done;
      (II) the mean power needed. [4]

(c) Assume the force $F$ calculated in (a) is now applied horizontally as shown. Calculate the initial acceleration of the sledge, given that its mass is 40.0 kg and assuming that the frictional force stays the same. [3]

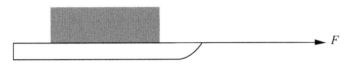

$F$

## Seren's answer

(a) $\cos 40 = \dfrac{200}{F}$ ✓

   $F = 261\ N$ ✓

(b) (i) Work done = $Fx\cos\theta$ ✓✓
   Force applied in the desired direction. As no vertical motion is desired ✗ it is not work done.

   (ii) (I) Work done = $Fx\cos\theta$
             = 200 N × 2000 m = 400 000 J ✓✓

      (II) 30 minutes = 1800 s ✓

      $P = \dfrac{E}{t} = \dfrac{400000}{1800} = 222\ W$ ✓

(c) $a = \dfrac{\Sigma F}{m}$, $\Sigma F = 261\ N - 200\ N$ ✓ $= 61\ N$

   $a = \dfrac{61}{40}$ ✓ $= 1.5\ m\ s^{-1}$. ✗ (unit penalty)

## Examiner commentary

(a) Clearly set out – both marks

(b) (i) Seren's statement, $Fx\cos\theta$, is accepted for 2 marks – the 1st mark is for force × distance and the 2nd for '. . . moved in the direction of the force' – the $\cos\theta$ is treated as equivalent to the 2nd statement.

   Strange expression: 'desired' is unclear and not accepted.

   (ii) (I) The 1st mark is a conversion mark [2.0 km → 2000 m] and the second for the correct multiplication.

      (II) Again, Seren has correctly converted 30 minutes to seconds and followed this by a correct application of the power equation.

(c) Seren has calculated the resultant force [1st mark] and correctly applied $a = \dfrac{\Sigma F}{m}$ [2nd mark]. She has given an incorrect unit for acceleration and has missed the 3rd mark.

**Seren gains 10 out of 12 marks.**

## Tom's answer

(a) $\cos 40 = \dfrac{200}{F}$ ✓

   $F = 200\cos 40 = 153\ \text{N}$ ✗

(b) (i) Work done = Force × distance moved ✓ ✗
       There is no work done vertically because the sledge doesn't move vertically, only horizontally. ✓

   (ii) (I)   2000 m ✓ × 200 N = 40 000 J ✗
        (II)  60 × 60 = 3600 (1 minute)
              3600 × 30 = 108 000 (30 minutes) ✗

              $P = \dfrac{W}{t} = \dfrac{40\ 000}{108\ 000}$ ✓ e.c.f. = 0.37 W

(c) $F = ma$
   $153 = 40 \times a$ ✗

   $\dfrac{153}{40} = a = 3.8\ \text{m s}^{-2}$ ✗ no e.c.f.

## Examiner commentary

(a) Tom has correctly substituted into the equation, gaining the 1st mark, but has re-arranged incorrectly and so loses the 2nd mark.

(b) (i) The definition of work done is incomplete because there is no statement about direction of motion. The subsequent statement is insufficient to overcome this deficit.
       Tom gains the 3rd mark, however, because he correctly identifies that there is no vertical motion.

   (ii) (I)  Tom correctly converts 2.0 km to m but makes a slip in multiplying.
        (II) Tom loses the 1st mark for the minute → second conversion but is allowed e.c.f. on both this and the incorrect work in the calculation of power.

(c) In this last part of the question, the examiners only gave any credit if candidates attempted to calculate the resultant force. As Tom didn't do this he received 0 for part (c).

**Tom gains 5 out of 12 marks.**

When 2 protons are accelerated to high kinetic energies and collide with each other, the following reaction may occur. [x is an 'unknown' particle.]

$$p + p \rightarrow p + x + \pi^{+}$$

(a) The charge on a proton (p) is $+e$.
   (i) What is the magnitude of the charge on the $\pi^{+}$ (a pion or $\pi$ meson)?  [1]
   (ii) Determine the charge of particle x.  [1]
(b) The $\pi^{+}$ is classed as a *meson*. How is p classed?  [1]
(c) In the reaction, u quark number and d quark number are each conserved. [$\bar{u}$ is assigned a u quark number of –1, and $\bar{d}$ a d quark number of –1.]
   Giving your reasoning, determine the quark make-up of particle x and hence identify this particle.  [4]
(d) Explain how *lepton* conservation applies in this reaction.  [1]
(e) Discuss which of the forces, *weak*, *strong* or *electromagnetic*, is likely to be responsible for the reaction.  [2]

## Seren's answer

(a) (i) +1, +e ✓
    (ii) 0e, neutral ✓
(b) baryon ✓
(c) quark formula: $uud + uud \rightarrow uud + x + u\bar{d}$
    ✓ for the protons, ✓ for the pion
    so quark make up of x is udd. ✓ The particle
    is a neutron. ✓
(d) The overall lepton number of each side is 0
    as no leptons are present. ✓
(e) The strong force ✓ is likely to be
    responsible because quarks are the only
    particles present in the reaction – there
    are no leptons. ✗

## Examiner commentary

(c) Reasonably full working – Seren has clearly
    given the quark make-up of the protons and
    pion and identified the resulting particle and
    its composition.
(d) This is a much clearer statement than Tom's.
(e) The second sentence doesn't quite pin it
    down because weak interactions involving
    only quarks are possible – but only when
    individual quark flavours are not conserved.

**Seren gains 9 out of 10 marks.**

## Tom's answer

(a) (i) +1 ✓
    (ii) 0 ✓
(b) quarks ✗
(c) ↑↑↓ + ↑↑↓ → ↑↑↓ + ↑↓↓
    up, down, down ✓ with a charge of 0 is a neutron ✓
(d) There is always 0 ✗
(e) Strong force ✓
    Because protons and neutrons are present in the product formula
    which can form an attractive or repulsive force between them. ✗

## Examiner commentary

(a) (i) Either +1 or +e was accepted.
    (ii) No explanation is required on this occasion. 0, 0e or
         neutral were accepted.
(b) The particle contains quarks – it is classed as a baryon.
(c) The first line was not credited – the notation clearly represents
    up and down quarks but the meson is missing and it is
    not clear how it relates to the answers, which are correct.
    'Giving your reasoning' means that the examiner needs to be
    convinced.
(d) It is not clear what this means.
(e) The second sentence may mean something, but again it is not
    clear.

**Tom gains 5 out of 10 marks.**

a)  (i) The spectrum of the star Rigel in the constellation of Orion peaks at a
        wavelength of 260 nm. Calculate the temperature of the surface of Rigel.     [2]
    (ii) What assumptions were you making about the way the star's surface radiates? [1]

(b) To a good approximation, the kelvin temperature of Rigel's surface is twice that of
    the Sun, and the radius of Rigel is 70 times the radius of the Sun. Use Stefan's Law
    to estimate the ratio:

$$\frac{\text{total power of electromagnetic radiation emitted by Rigel}}{\text{total power of electromagnetic radiation emitted by the Sun}}$$     [3]

(c) We can discover the presence of particular atoms in the atmosphere of a star by
    measuring the wavelengths of dark lines in the star's spectrum.
    Explain how the lines arise, and why they occur at specific wavelengths.     [3]

## Seren's answer

(a) (i) $\lambda_{max} T = W$

$250 \times 10^{-9} \times T = 2.90 \times 10^{-3}$ ✓

$T = 11154° \times$ (unit)

(ii) It radiates as a black body. ✓

(b) Rigel $r = 70$ Sun $= 1$

$P_R = 4\pi \times 70^2 \times 5.67 \times 10^{-8} \times 11154^4 = 5.402 \times 10^{13}$

$P_S = 4\pi \times 1^2 \times 5.67 \times 10^{-8} \times (\frac{1}{2} \times 11154)^4 = 6.89 \times 10^8$ ✓✓

Ratio $= \dfrac{5.40 \times 10^{13}}{6.89 \times 10^8}$, so Rigel $= 78374$ times more

power than the Sun ✓

(c) Radiation of all wavelengths is emitted from the star. The photons emitted pass through the atmosphere of the star. ✓ If a photon collides with an electron and the photon's energy coincides with an energy gap jump for the electron, the photon is absorbed ✓ by the electron jumping to a higher energy level. This shows up as a dark vertical line on the coloured spectrum at the wavelength absorbed. ✓

---

### Examiner commentary

(a) (i) Seren correctly calculates the kelvin value of the temperature. Unfortunately she omits the unit – ° is not a unit of temperature.

(b) Seren uses the relative values of radius, and correctly squares this ratio. She raises the temperatures to the 4th power. She calculates the two powers – the fact that she does not use actual radii and hence does not express the powers in SI units does not matter: she correctly finds the ratio. Not a pretty answer but…

(c) Seren correctly identifies:

- that the radiation with a continuous spectrum passes through the atmosphere of a star;
- absorption happens by interaction with atoms with a specific energy gap
- the dark line corresponds to the energy of the absorbed photons.

**Seren scores 8 out of 9 marks**

## Tom's answer

(a) (i) $\lambda_{max} = WT^{-1}$

$250 \times 10^{-9} = 2.90 \times 10^{-3} T^{-1}$ ✓

$T^{-1} = 8.97 \times 10^{-5}.\quad T = 1.13 \times 10^4$ K ✓

(ii) Temperature is the same over the surface ✗

(b) $T_R = 2T_S; R_R = 70R_S$

$P = A\sigma T^4$.  $\dfrac{P_R}{P_S} = \dfrac{4\pi r^2 \times \sigma \times T^4}{4\pi r^2 \times \sigma \times T^4} = \dfrac{r^2 \times T^4}{r^2 \times T^4}$ [$\pi$ and $\sigma$ are the same]

$= \dfrac{1^2 \times 11200^4}{70^2 \times \left(\dfrac{11200}{2}\right)^4}$ ✓ ✗

$= \dfrac{4}{1125} = 3.27 \times 10^{-3}$

(c) A continuous spectrum is given off from a star but an absorption spectrum is seen when looking at the star. This is because gases absorb certain wavelengths of light. ✓ ∴ we can discover what gas a star is made up from.

---

### Examiner commentary

(a) (i) Tom selects the correct equation and substitutes data correctly, earning the first mark. He correctly calculates the temperature and gives the unit, earning the second mark.

(ii) Tom misses the point that Wien's Law is only true for black bodies.

(b) Tom correctly expresses Stefan's Law to obtain the ratio. He correctly handles the 4th power of the temperatures and gains a mark but unfortunately inverts the ratio of the square of the radii. Thus his final answer is incorrect. Tom's initial statement, $T_R = 2T_S$, is correct but he does not use it.

(c) To obtain more than one mark, Tom needs to discuss the absorption of photons / atomic energy levels and also to indicate where the absorption occurs.

**Tom scores 4 out of 9 marks**

**Q & A 6**

(a) Derive, giving a labelled diagram, the relationship between the current $I$ through a metal wire of cross-sectional area $A$, the drift velocity, $v$, of the free electrons, each of charge $e$, and the number, $n$, of free electrons per unit volume of the metal. [4]

($I = nAve$)

(b) Calculate the drift velocity of free electrons in a copper wire of cross-sectional area $1.7 \times 10^{-6}$ m$^2$ when a current of 2.0 A flows. [$n_{copper} = 1.0 \times 10^{29}$ m$^{-3}$]. [2]

(c) A potential difference is required across the copper in order for the current to flow. The size of the current depends upon the wire's *resistance*. Explain in terms of free electrons, how this resistance arises. [2]

(d) The copper wire in (b) is of length 2.5 m. When it carries a current of 2.0 A it dissipates energy at the rate of 0.1 W. Calculate its resistivity. [4]

(e) A second copper wire has the same volume as the wire in (d) but is longer. Complete the table below indicating whether the quantity given is **bigger**, **smaller** or **the same** for this longer wire. [3]

| Quantity | For the longer wire this quantity is: |
|---|---|
| Cross-sectional area | |
| $n$, the number of free electrons / unit volume | |
| Resistivity | |

## Seren's answer

(a)
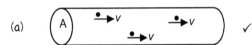
✓

$Q$ = number of electrons × charge on each electron
$Q = V \times n \times e$ ✓
since $V = AL$    $Q = ALne$ ✓

$I = \dfrac{Q}{t}$  ∴ $I = \dfrac{ALne}{t}$

since $v = \dfrac{L}{t}$    $I = nAve$ ✓

(b) $v = \dfrac{I}{nAe}$     $v = \dfrac{2.0}{1.0 \times 10^{29} \times 1.7 \times 10^{-6} \times 1.6 \times 10^{-19}}$ ✓

$v = 7.4 \times 10^{-5}$ m s$^{-1}$ ✓

(c) The free electrons that travel in the wire collide ✓ with ions ✓, which give its resistance. If, for example, the copper is heated, it means the ions vibrate more in a sphere of influence, and more electrons collide with them, giving higher resistance.

(d) $\rho = \dfrac{RA}{L}$.       $P = I^2R$ ✓ ∴ $R = \dfrac{P}{I^2} = \dfrac{0.1}{2.0^2} = 0.025$ Ω. ✓

$\rho = \dfrac{0.025 \times 1.7 \times 10^{-6}}{2.5}$ ✓ $= 1.7 \times 10^{-8}$ Ω m$^{-1}$ ✓ [note incorrect unit]

(e) same     ✗
same     ✓
smaller     ✗

## Examiner commentary

(a) Not a perfect derivation; for example, $I$ is not defined, and there is no commentary, but the answer hits all the marking points.

(b) First mark for substitution into the equation – Seren has manipulated the equation too, which was not necessary for this mark – and the second for the correct answer.

(c) Seren has answered the question in the first sentence. The second sentence answers a different question.

(d) Seren identifies the resistivity equation, realises she needs first to find the resistance, which she correctly does using $P = I^2R$, and then goes on to calculate the resistivity. She uses the wrong unit for resistivity but is lucky that, on this occasion, there was no unit penalty!

(e) Seren has not realised that, if the volume of the longer wire is the same as that of the shorter wire, its cross sectional area must be less! The resistivity is a characteristic of the <u>material</u> and does not depend upon the shape of the specimen.

**Seren gains 13 out of 15 marks.**

# Tom's answer

(a)

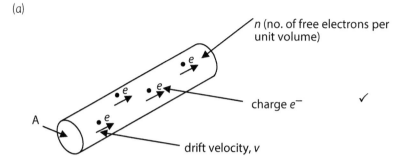

$n$ (no. of free electrons per unit volume)

charge $e^-$ ✓

drift velocity, $v$

A

$I = nAve$ for current      velocity $= \dfrac{I}{nAe}$      $n = \dfrac{I}{Ave}$

$A = \dfrac{I}{nev}$      $e = \dfrac{I}{Ane}$

(b) $v = \dfrac{I}{nAe}$

$= \dfrac{2.0}{1.7 \times 10^{-6} \times 1.0 \times 10^{-16} \times 1.0 \times 10^{29}}$ ✗ – incorrect substitution

$= 1.176 \times 10^{-7} \text{ m s}^{-1}$ ✗ – no e.c.f.

(c) Free electrons in the wire knock into ✓ ions in the metal lattice ✓ they block the way and slow electrons down giving resistance.

(d) $R = \dfrac{V}{I}$.      $P = VI$    so $R = \dfrac{I^2}{P} = \dfrac{4}{0.1} = 40 \ \Omega$

$\rho = \dfrac{RA}{I} = \dfrac{40 \times 1.7 \times 10^{-6}}{2.5}$ ✓ (e.c.f.) $= 2.72 \times 10^{-5} \ \Omega \text{ m}$ ✓

(e) smaller    ✓
same      ✓
smaller    ✗

## Examiner commentary

(a) Good diagram – excessive labelling as the quantities are defined in the question stem. Tom clearly has not learnt the derivation. He has just played with the equation.

(b) Tom has made a mistake with his substitution – the value for $e$ is wrong. He has not helped himself by not keeping the quantities on the bottom line of the fraction in the same order as that in the algebraic equation. The second mark cannot be given following an incorrect substitution.

(c) Tom has given a satisfactory answer here.

(d) Tom has made a mistake in finding $R$. His use of this incorrect value was credited on the e.c.f. principle and therefore so was his incorrect final answer.

(e) The resistivity of a metal does not depend upon its shape – only the composition.

**Tom gains 7 out of 15 marks.**

**Q&A 7**

> (a) What is a superconductor? [1]
> (b) With the aid of a sketch graph, explain the term superconducting transition temperature. [3]
> (c) Explain why superconductors are useful for applications which require large electric currents and name **one** such application. [2]

## Seren's answer

(a) A material that, beyond a certain temperature, its resistance falls effectively to 0. ✓bod

(b)

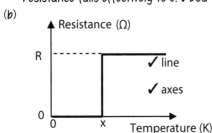

**X** is the temperature beyond which the resistance effectively becomes 0. A transition between resistance R and 0 occurs at this temperature. ✓

(c) A large amount of current can pass through them easily (i.e. they will feel no resistance). They are used in particle accelerators. ✓

### Examiner commentary

(a) On this occasion the examiner was only looking for the 0 resistance – the idea of transition temperature is covered in (b).

(b) Seren has correctly identified the axes and given a correct graph. 'Beyond' is not a good word to use, but it is clear from the graph that she means 'below'.

(c) The missing mark is because no link was made to 0 energy loss with a superconductor. Particle accelerators was accepted for the mark but ideally Seren would have identified the magnets as requiring the superconductors.

**Seren gains 5 out of 6 marks.**

## Tom's answer

(a) A conductor with zero resistance. ✓

(b)

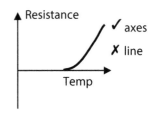

This is the temperature at which all resistance disappears. ✗

(c) Power cables – less energy is wasted through friction/heat ✓ ∴ they are more efficient.

### Examiner commentary

(a) Tom hits the marking point.

(b) Tom has correctly identified that the significant graph is one of resistance against temperature but has drawn an inadequate line and his description does not make it clear that the 0 resistance covers a range of temperatures from 0 to the transition temperature.

(c) Tom identifies the advantage as involving energy loss ['friction' is ignored here] but it was not accepted that power cables use superconductors.

**Tom gains 3 out of 6 marks.**

A student directs a narrow beam of light on to one end of a glass block, as shown:

(a)  (i)  Referring to the diagram, calculate the angle of incidence, $x$. [Refractive index of air = 1.00; refractive index of the glass = 1.52]   [3]
     (ii)  Calculate the angle $y$.   [1]
     (iii) Show that the light does not refract into the air at point **P**.   [2]
     (iv)  (I) The light changes its direction of travel at point **P**. What is the full name for the process involved?   [1]
           (II) How does the size of the angle $z$ compare with the size of angle $y$?   [1]
(b)  (i)  A glass fibre used for the transmission of data consists of a central glass core with a *cladding* of glass of lower refractive index. Suggest one advantage of having glass cladding rather than simply an air surround.   [1]
     (ii)  What can be said about the diameter of a monomode fibre?   [2]
     (iii) Why is such a fibre called monomode?   [1]

## Seren's answer

(a)  (i)  $n_1 \sin\theta_1 = n_2 \sin\theta_2$

   $\therefore 1 \sin x = 1.52 \sin 25$ ✓;  $\therefore \sin x = 0.6423...$

   $\therefore x = \sin^{-1} 0.642$ ✓ $= 40.0°$ (to 3.s.f.)✓

 (ii)  $180 - (25 + 90) = 65°$✓

 (iii)  $n_1 \sin\theta_1 = n_2 \sin 90$

   $\therefore 1.52 \sin\theta_1 = n_2$ ✓  $\therefore$ critical angle $= \sin\left(\dfrac{1}{1.52}\right) = 41.1°$

   $\therefore 65° >$ critical angle $(41.1°)$ ✓

   $\therefore$ light does not refract

 (iv)  (I) Total internal reflection. ✓

   (II) $z = y$ ✓

(b)  (i)  It reduces the critical angle, thereby reducing the amount of paths. ✗ not bod

 (ii)  It is very small ✓

 (iii)  It only allows one path for transmission. ✓

## Examiner commentary

(a)  (i)  The first mark is for putting correct data into the equation; the second for the answer.

 (iii)  Seren has applied the correct principle: 1st mark – correct insertion into the equation; 2nd mark for the correct discussion.

(b)  (i)  This is a statement in the right area [effect on critical angle] but it is the wrong way round (cladding will <u>increase</u> the critical angle) and there is no clear link to the suggested advantage, i.e. reduction of multimode dispersion.

 (ii)  This is the first marking point. The second required an approximate size [less than ~ 1 μm was accepted] or a comment about the size in relation to the wavelength of the radiation used.

 (iii)  A rare correct statement!

**Seren gains 10 out of 12 marks.**

## Tom's answer

(a) (i) $n = \dfrac{\sin i}{\sin r} = \dfrac{\sin 1}{\sin 1.52}$ ✗ $= 0.647$   $\sin^{-1} = 41°$

(ii) $130 \div 2 = 65°$ ✓

(iii) Because for the light to refract out into the air, angle x must be smaller for the light to not refract inwards but go outwards instead. ✗

(iv) (I) Total internal refraction ✗

(II) They are the same size. ✓

(b) (i) It means that only beams with small angles of incidence stay inside the fibre, so all the beams inside arrive at their destination at similar times.✓

(ii) Very small ✓

(iii) Only one bit of information can be passed down it. ✗

## Examiner commentary

(a) (i) Tom has selected a version of the correct formula but has missed the first mark because he has inserted incorrect data. No mark can be given for the answer resulting from this.

(ii) The answer is correct: the method is slightly obscure but not demonstrably incorrect.

(iii) Tom fails to realise that a calculation is needed here. The comment about x is irrelevant as it is the angle y which is significant.

(iv) (I) Tom has used the word 'refraction' instead of 'reflection' – this confusion also occurs in the answer to (a) (iii) – no e.c.f. is allowed for this mistake of principle.

(b) (i) A good answer from a weak student.

(ii) As Seren's answer.

(iii) A common misunderstanding.

**Tom gains 4 out of 12 marks.**

Q&A 9

(a) (i) What is the *photoelectric effect*? [2]

(ii) Give an account of the photoelectric effect in terms of photons, electrons and energy, explaining how it leads to *Einstein's photoelectric equation*. [4]

(b) (i) A zinc surface of work function $4.97 \times 10^{-19}$ J is irradiated with two frequencies of electromagnetic radiation in turn. For each frequency, show whether or not electrons are emitted from the surface and, if they are emitted, calculate their maximum kinetic energy.

(I) $7.99 \times 10^{14}$ Hz. [2]

(II) $6.74 \times 10^{14}$ Hz [1]

(ii) What would be the maximum kinetic energy of the electrons emitted if the surface were irradiated with both frequencies at once? Explain your reasoning. [2]

## Seren's answer

(a) (i) The emission of an electron from the metal surface ✓ with the energy provided by photon energy of light source. ✓

(ii) The energy of a photon required to liberate ✓ an electron should be equal to either the work function (minimum) ✓ or more (work function = minimum energy to liberate an electron). For photons with greater energy than the work function, the kinetic energy of the liberated electron is equal to the difference between $E_{photon}$ and $\phi$, ✓ i.e. $KE_{max} = hf - \phi$, where $hf$ = photon energy ✓ and $\phi$ = work function.

(b) (i) (I) Photon energy = $6.63 \times 10^{-34} \times 7.99 \times 10^{14} = 5.297 \times 10^{-19}$

It is bigger than $\phi$ and so electrons are emitted. ✓

Max KE = $5.297 \times 10^{-19} - 4.97 \times 10^{-14} = 3.27 \times 10^{-20}$ J ✓

(II) Photon energy  $= 6.63 \times 10^{-34} \times 6.74 \times 10^{14}$

$= 4.47 \times 10^{-19} < \phi$

So no electron is emitted. ✓

(ii) $3.27 \times 10^{-20}$ J, because the metal will absorb the light with the higher frequency for emitting electrons. ✓✓

### Examiner commentary

(a) (i) Electrons from metal surface; energy provided by photons – 2 clear marks

(ii) The account is rambling, but all the marking points are covered: photon energy; work function; indication that a single photon liberates a single electron; clear statement of energy housekeeping.

(b) (i) (I) Seren's statement is different from Tom's and equally valid.

(II) Seren has shown clearly that the photon has insufficient energy and drawn the correct conclusion.

(ii) Seren's answer is not perfect, but enough for the mark: she has correctly given the energy; a statement that the photon energies cannot combine, or similar, would have been better.

**Seren gains 11 out of 11 marks.**

## Tom's answer

(a) (i) When photons of the correct energy are absorbed and electrons are emitted. ✗✓

(ii) Einstein's photo-electric equation: $\frac{1}{2}mv^2 = hf - \phi$

$\frac{1}{2}mv^2$ is the maximum kinetic energy of electron

$hf$ is the frequency of photon. ✗

$\phi$ is the minimum energy needed to remove the electron from the surface ✓

– sign is the energy escape left.

(b) (i) (I) $E = hf - \phi = 6.63 \times 10^{-34} \times 7.99 \times 10^{14} - 4.97 \times 10^{-14}$ ✓

$= 3.27 \times 10^{-20}$ J ∴ Electrons are emitted. ✓

(II) $E = hf - \phi = 6.63 \times 10^{-34} \times 6.74 \times 10^{14} - 4.97 \times 10^{-14}$

$= -5.01 \times 10^{-20}$ J ✗

(ii) It will be equal to zero as the frequency is minimum. ✗

### Examiner commentary

(a) (i) The absorption/emission mark is awarded, but not the context.

(ii) It's a good idea to state the equation but the derivation should be clear. The identification of $\phi$ is clear and gains a mark. Tom calls $hf$ the 'frequency' of the photon – 'energy' would have gained the extra mark.

Unfortunately the mark scheme awarded no mark for $\frac{1}{2}mv^2$ and the last statement needs too much interpretation to credit.

(b) (i) (I) Good – clear and concise. Tom derives the maximum energy of the emitted electrons and states that they are emitted.

(II) There is no mark just for repeating a calculation. Tom should have drawn attention to the – sign and indicated that no electrons would be emitted.

(ii) This statement has no worth.

**Tom gains 4 out of 11 marks.**

In the helium-neon laser, excited helium atoms collide with neon atoms and transfer energy to them. This raises neon atoms from the ground state to the excited *metastable* state, **U** (see diagram).

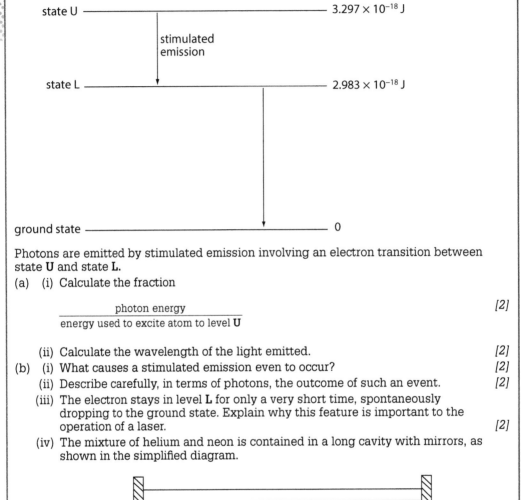

Photons are emitted by stimulated emission involving an electron transition between state **U** and state **L**.

(a) (i) Calculate the fraction

$$\frac{\text{photon energy}}{\text{energy used to excite atom to level } \mathbf{U}}$$

[2]

    (ii) Calculate the wavelength of the light emitted. [2]

(b) (i) What causes a stimulated emission even to occur? [2]

    (ii) Describe carefully, in terms of photons, the outcome of such an event. [2]

    (iii) The electron stays in level **L** for only a very short time, spontaneously dropping to the ground state. Explain why this feature is important to the operation of a laser. [2]

    (iv) The mixture of helium and neon is contained in a long cavity with mirrors, as shown in the simplified diagram.

mirror reflecting almost 100% of light     *amplifying medium* of helium and neon     mirror reflecting about 99% of light

How does this cavity design promote laser operation? [2]

## Seren's answer

(a) (i) photon energy $= 3.297 \times 10^{-18} - 2.983 \times 10^{-18}$ ✓

$= 3.14 \times 10^{-19}$ J

$\therefore$ Fraction $= \dfrac{3.14 \times 10^{-19}}{3.297 \times 10^{-18}} = \dfrac{2}{21}$ ✓

(ii) $E_{photon} = \dfrac{hc}{\lambda} \quad \therefore \lambda = \dfrac{hc}{E_{photon}}$ ✓ $= \dfrac{6.6 \times 10^{-34} \times 3 \times 10^{8}}{3.14 \times 10^{-19}}$

$= 6.31 \times 10^{-7}$ m ✓

(b) (i) An incident photon on an atom ✓ causes it to rise to an excited state then quickly drop down, releasing a photon, which causes a knock-on effect on other excited states ✗

(ii) The knock on effect of photons being released at practically the same time causes all photons to be coherent ✓ with each other causing light amplification (which is monochromatic).

(iii) It makes a population inversion easier to achieve. ✓ If there was no population inversion, then there will be more absorbtion than emission ✓ and the amplifying medium will not function.

(iv) It allows photons to move back and forth within the amplifying medium ✓ and $\therefore$ increase in intensity ✓ while allowing 1% of light out, to carry out the function of the particular laser.

## Tom's answer

(a) (i) $\dfrac{3.297}{(3.297 - 2.983)} = \dfrac{21}{2}$ ✗

(ii) $E = hf, E = h\left(\dfrac{c}{\lambda}\right)$

$\therefore 3.14 \times 10^{-19} = 6.63 \times 10^{-34} \times \left(\dfrac{3 \times 10^{8}}{\lambda}\right)$

$\lambda = 6.33 \times 10^{-7}$ m ✓✓

(b) (i) A photon hitting the electron encouraging it to drop to a lower energy. ✓

(ii) The photon will be of the same frequency ✓ as the one that hit it and they ✓ will be in phase.

(iii) The photon emitted does not cause stimulated emission but spontaneous emission, which will not cause coherent photons to be emitted. ✗

(iv) Most photons produced will reflect back and stimulate more with the same frequency. ✓

## Examiner commentary

(a) (i) Clearly laid out: the desired answer was 0.095 but most candidates gave $\dfrac{2}{21}$ and this was accepted.

(ii) Correct equation produced for $\lambda$ (1st mark); correct answer (2nd mark).

(b) (i) Seren has identified the event as being caused by an incident photon but has mistakenly stated that this causes excitation.

(ii) Seren's answer is not worth two marks because it fails to deal with a single event, i.e. a single photon incident leading to 2 photons emerging.

(iii) Seren's is an excellent response. The spelling mistake, even in a significant word, is ignored.

(iv) Seren's answer picks out both marking points clearly – see Tom's answer and Examiner commentary.

**Seren gains 10 out of 12 marks.**

## Examiner commentary

(a) (i) How unfortunate! Tom has noticed that the factor of $10^{-18}$ is common, so can be omitted, but has inverted the fraction. No credit gained.

(ii) The first mark is for combining the equations $E = hf$ and $c = f\lambda$ and manipulating to produce an equation with $\lambda$ as the subject; the second mark for the answer.

(b) (i) Tom has, somewhat clumsily, identified an event as being caused by an incident photon. He has missed the point that the photon's energy must equal the energy gap U–L.

(ii) Tom has identified that the stimulated photon will have the same frequency as the incoming one [better would have been 'coherent']. His comment that 'they will be in phase' implies that there are two photons where previously there was one.

(iii) Tom does not address the importance of population inversion.

(iv) Tom has partly addressed two marking points (multiple traverses of the cavity; increasing the chances of stimulated emission) and is credited with one mark.

**Tom gains 6 out of 12 marks.**

An experiment is carried out to determine the resistivity of a metal using variable lengths of the metal in the form of a wire. Explain how the experiment should be carried out and how an accurate value of the resistivity could be obtained from the results. [6 QER]

## Seren's answer

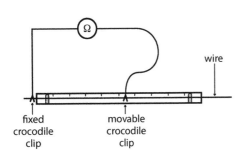

fixed crocodile clip

movable crocodile clip

wire

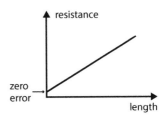

The wire is taped to a metre rule as shown. By clipping the movable crocodile clip to at least 10 different positions, a series of readings of resistance, as measured on the ohm-meter, and length, measured on the metre-rule scale, is taken. The diameter of the wire is measured using a micrometer or pair of digital callipers, so that the cross-sectional area can be calculated.

A graph of resistance against length is drawn and appears as shown. The zero error, $R_0$, is the resistance of the ohm-meter leads. The equation of the graph is $R = \frac{\rho l}{A} + R_0$, where $l$ is the length and A the cross sectional area of the wire. The gradient of the graph is $\frac{\rho}{A}$, so the resistivity is the gradient × the cross sectional area.

---

### Examiner commentary

**Diagram:** The diagram shows a correct set-up and will allow any student to replicate the experiment. Minor faults are that neither the metre rule nor the taping points of the wire are labelled.

**Method:** Seren clearly details the measurements that need to be taken. She doesn't mention taking repeat readings, which is necessary for the diameter of the wire. Arguably repeat readings are not needed for resistance – just a large number of readings. She also doesn't mention finding the zero error but does deal with this in his treatment of the results.

**Results:** Plotting a graph of resistance against length and quoting the equation $R = \frac{\rho l}{A}$ together with the treatment of the zero error are well described. Seren correctly identifies how the resistivity will be calculated from the gradient of the graph. She doesn't actually mention finding the gradient but that is strongly implied. A more serious failing is that she does not explain how the cross sectional area of the wire is determined.

**Conclusion:** This is an upper band answer but with a few omissions. An examiner would probably give her 5 marks out of 6.

## Tom's answer

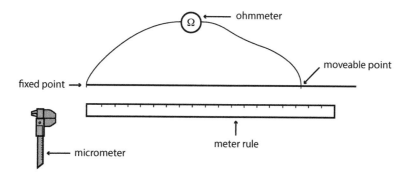

Use the meter ruler to measure the length(s) of the wire and the micrometer to measure the diameter and therefore the cross-sectional area of the wire. Take note of the resistance of the various lengths and then use the three results of resistance, area and length of the wire in the equation $\rho = \dfrac{RA}{l}$ to get an answer for the resistivity.

### Examiner commentary

**Diagram**: Tom has drawn a circuit that will work and identified method of changing the lengths of wire within it. He has, by implication indicated how he will measure the length of the wire. His inclusion of the micrometer is unnecessary: it takes time to draw, it is not shown in position and it is mentioned in the written text.

**Method**: The first sentence of Tom's paragraph contains all his description of the method of obtaining results. It is barely adequate. To obtain high marks the examiner would have expected to see repeat readings of the diameter at different points along the wire and possibly recording the zero error in resistance (i.e. the resistance with the two ohm-meter leads connected together).

**Results**: The only marking point here which Tom has hit is the use of the equation, $\rho = \dfrac{RA}{l}$, for resistivity. He mentions using three results but doesn't say how he will use them, e.g. calculate $\rho$ for each set of results and then finding the mean value. He doesn't say how he will calculate the cross-sectional area (using $A = \pi r^2$). Most seriously, he doesn't aim to draw a graph of resistance against length and use the gradient in calculating $\rho$.

**Conclusion**: Tom's answer is a middle band answer and would expect to attract about half marks.

# Quickfire answers

## Section 1.1

① $[ut] = [u][t] = $ m s$^{-1}$ × s = m
$[\frac{1}{2}at^2] = [a][t^2] = $ m s$^{-2}$ × s$^2$ = m
∴ RHS is homogeneous
[LHS] = [x] = m = [RHS] ∴
equation is homogenous. QED

② kg m$^{-1}$ s$^{-2}$

③ J = [Fx] = kg m s$^{-2}$ × m = kg m$^2$ s$^{-2}$.
QED
$W = \dfrac{J}{s} = $ kg m$^2$ s$^{-2}$ × s$^{-1}$ = kg m$^2$ s$^{-3}$.

④ kg m$^{-1}$ or N m$^{-1}$ s$^2$

⑤ 56 μm

⑥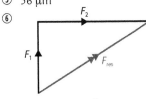

⑦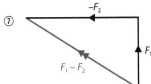

⑧ 50 m s$^{-1}$ at 38.9° E of S

⑨ 19.3 kN

⑩ 1.73 m s$^{-2}$ horizontally to the right

⑪ 47.4 kg

⑫ (a) $2.0F_1 = 0.5F_2 + 2.0 × 500$
(b) $4.0F_1 + 0.5 × 75 = 2.5F_2$
(c) As in the example

⑬ 3.8°

⑭ (a) 0.753 g cm$^{-3}$
(b) 0.753 ± 0.010 g cm$^{-3}$ or
0.75 ± 0.01 g cm$^{-3}$

⑮ (a) $x$ = 17.6 ± 0.2 cm;
$y$ = 25.60 ± 0.15 cm
(b) $M$ = 44.3 ± 0.8 g

## Section 1.2

① 41 m s$^{-1}$ NW

② (a) 45 m s$^{-1}$ N
(b) 45 m s$^{-1}$ SW

③ Zero
(because total displacement = 0)

④ Instantaneously at $t$ = 0 and from 8.0–10.0 s.

⑤ ....Negative acceleration (deceleration) to rest between 2.4 and 4.0 s.
Stationary 4.0–5.0 s followed by reverse direction and accelerating in a negative direction up to 5.8 s and carrying on at a steady (negative) velocity to 7.2 s (Passes point of origin at 6.9 s). Then slowing down (positive acceleration because the velocity is negative) to rest at 8.0 s and remaining at rest until 10.0 s.

⑥ (a) 0–5 s and 22.5–25 s
(b) 0–5 s and 15–18 s
(c) 5–11.5 s and 18–22.5 s
(d) 11.5–18 s
(e) 11.5–15 s and 22.5–25 s

⑦ 5.0 m s$^{-2}$, 0, – 7.1 m s$^{-2}$, –2.0 m s$^{-2}$, 0, 1.2 m s$^{-2}$

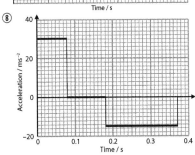

⑧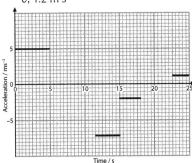
Displacement = 0.71 m

⑨ Algebra gives ± 17 m s$^{-1}$. The speed must be positive, so 17 m s$^{-1}$ is correct.

⑩ (a) 0.84 s and 3.65 s. One time on the way up and one on the way down.
(b) Because the greatest height reached is 24.4 m.

⑪ 0.7 m

⑫ Time method:
$v_y = \sqrt{2gy} = $ 31.3 m s$^{-1}$
Drop distance method:
$v_d = \sqrt{2gy} = \sqrt{2 × 9.81 × 50}$
= 31.3 m s$^{-1}$
Resultant velocity = 43.4 m s$^{-1}$ at 46° below the horizontal.

⑬ 1. As in the example
2. $v_y$ = −9.15 m s$^{-1}$
3. $t$ = 1.61 s
4. As in the example

## Section 1.3

① Weight – the gravitational force of the plane on the Earth
Air resistance – the forward drag force of the plane on the air

②
| Force acting | N3 partner |
|---|---|
| Weight | gravitational force of the skydiver on the Earth |
| Air resistance | drag of the skydiver on the air |

③ Because the skydiver is not accelerating, we know that the weight and air resistance are equal and opposite. They are not an N3 pair because they act on the same body; they are also different types of force.
The gravitational force of the skydiver on the Earth and the drag of the skydiver on the air are also equal and opposite (because each is equal to one of two forces that we know are equal). They are not an N3 pair because they are different types of force (gravitational and intermolecular); also, prior to terminal velocity they wouldn't have been equal.

④ Gravitational force of Sun on Earth; ditto of Earth on Sun.

⑤ Gravitational force of skydiver on Earth = 800 N upwards
Drag force of skydiver on the air = 100 N downwards

⑥ $N = [F] = [ma] = kg\,m\,s^{-2}$
∴ $N\,s = kg\,m\,s^{-1}$ QED

⑦ (a) 20 000 kg m s⁻¹
   [or 20 000 N s,
   or 20 kN s]
   (b) $4.98 \times 10^{-20}$ N s

⑧ The N3 partner is the force exerted by the water on the gun.
Free-body diagram:

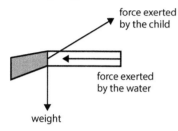

force exerted by the child

force exerted by the water

weight

⑨ (a) ~ 3100 N
   (b) 15 000 N
   (c) 2900 N

⑩ (a) 1.7 N s West
   (b) 110 N West

⑪ 110 N East

⑫ 1.6 m s⁻² to the left

⑬ 53.0 m s⁻² in a direction 23.5° upwards to the left.

⑭ 0.9 m s⁻¹ to the right (direction is obvious)

⑮ 8 m s⁻¹ to the right (again!)

⑯ 13.75 J (Too many s.f. really)

⑰ Recoil velocity = 5 m s⁻¹ (backwards)
   Total KE = 915 kJ
   Note that the KE of the gun is negligible (15 kJ which is 1.6% of the total). This is similar to the energy partition in α decay.

## Section 1.4

① 1260 m

② The Sun's gravitational force is at right angles to the motion of Venus. $F\cos 90° = 0$.

③ 9 MJ

④ 500 m s⁻¹

⑤ 7.8 g

⑥ 7.4 MJ

⑦ $x = 0.036$ m; $E_p = 56$ mJ

⑧ $E_k = \frac{1}{2}mv^2 = , \therefore v = \sqrt{\dfrac{2E_k}{m}}$

$= \sqrt{\dfrac{2 \times 0.192}{0.4}} = 0.98$ m s⁻¹

Assumption, no energy dissipated, e.g. by air resistance

⑨ The unit 'GW per year' is a rate of increase in power. 38.4 GW per year suggests that an additional 38.4 GW power station is switched on each year. A correct statement is 'the electrical power generated in 2012 was 38.4 GW'.

⑩ 17.5 TW h

⑪ As the submarine accelerates, the hydrodynamic resistance (water resistance) increases. When the resistance is equal to the thrust, all the work done by the engine is done against the resistance, so no additional kinetic energy is created, i.e. the submarine's speed stays constant.

⑫ 360 kW

## Section 1.5

① 19.6 N m⁻¹

② 23 cm

③ $[E] = kg\,m^{-1}\,s^{-2}$

④ $E = \dfrac{\sigma}{\varepsilon} = \dfrac{1}{\varepsilon} \times \sigma = \dfrac{l_0}{\Delta l} \times \dfrac{F}{A}$

$= \dfrac{Fl_0}{A\Delta l}$ QED

⑤ (a) 0.001 0 m (= 1.0 mm)
   (b) 0.001 0 km (= 1.0 m)
   (c) 0.53 mm

⑥ (a) 2500 N cm⁻²
   (b) $2.5 \times 10^7$ Pa
   (c) 25 MPa

⑦ 0.75 J

⑧ 200 kJ

⑨

⑩ Hammering hardens the copper because the dislocations are no longer free to move.

Heating and re-crystallising produces mobile dislocations and therefore softens the copper.

⑪ (a) UTS = 570 MPa
   (b) $\varepsilon = 0.0071$
   (c)

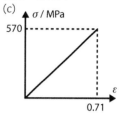

## Section 1.6

① (a) Visible; (b) infra-red;
   (c) infra-red

② It gives out all visible wavelengths but the peak emission is in the long-wavelength (yellow-orange) part of the spectrum.

③ (a) 4660 K
   (b) $1.26 \times 10^{28}$ W
   (c) $9.9 \times 10^{-9}$ W m⁻²

④ $\dfrac{r_{Arcturus}}{r_{Sun}} = 20, \therefore \dfrac{A_{Arcturus}}{A_{Sun}} = 400$

and $\dfrac{V_{Arcturus}}{V_{Sun}} = 8000$

⑤ $\dfrac{\rho_{Sun}}{\rho_{Arcturus}} = 7300$

⑥ uv

⑦ (a) $3.0 \times 10^{33}$ W
   (b) 88 million km

## Section 1.7

① Anti-strange quark $\bar{s}$.

② We know that there must be three quarks and the total charge must be zero.

$\frac{2}{3} + \left(-\frac{1}{3}\right) + \left(-\frac{1}{3}\right) = 0$

which is the only way of making zero. ∴ udd

③ $Q = -1$; antiproton or anti-deltaplus

④ $\bar{d}\bar{d}\bar{d}$

⑤ $\pi^+ = u\bar{d}$; $\pi^+ = \bar{u}d$
   So the u and anti-u quarks can annihilate as can the d and anti-d.

⑥ $B$   1 + –1   0 + 0 + 0
∴ $B$ is 0 on both sides
$Q$   1 + –1   2 + –2 + 0
∴ $Q$ is 0 on both sides
$L$   0 + 0   0 + 0 + 0
∴ $L$ is 0 on both sides

⑦ About $10^{-18}$ s because it is controlled by the electromagnetic interaction.

⑧ A neutrino is involved, which only feels the weak interaction
There is a change of quark flavour
The half-life of the decay is long (>5000 years for C-14)

⑨ $\Delta^+$: $U = 2$; $D = 1$
$p + \pi^0$: $U = 2$; $D = 1$
[$\pi^0$ has $U = D = 0$], i.e. same

⑩ (a) $\Delta^+$: → n + $\pi^+$
(b) uud → udd + u$\bar{\text{d}}$
[so, essentially, u → d + u$\bar{\text{d}}$]

⑪ (a) The e$^+$ and e$^-$ are leptons and therefore do not feel the strong force.
(b) All particles are charged and there is no change in quark numbers so could be controlled by either the e-m or weak interaction. The e-m interaction is much 'stronger' than the weak and therefore is responsible.

## Section 2.1

① Diameter: $1.0 \times 10^{-10}$ m
Nucleus: $1.0 \times 10^{-15}$ m
Electron: $1.0 \times 10^{-18}$ m

② –4.0 nC ($-4.0 \times 10^{-9}$ C)

③ 41.3 s.

④ 35 A

⑤ 33 s.w.g.

## Section 2.2

① 336 J

② (a) 5.9 V
(b) 5.76 C
(c) 34 J

③ (a) 13.0 A
(b) 17.6 $\Omega$

④ 44.1 $\Omega$

⑤ (a) $I = \dfrac{RA}{\rho}$
(b) $A = \dfrac{\rho l}{R}$

⑥ 17.3 $\Omega$, assuming a linear variation of resistance with temperature

## Section 2.3

① The currents into and out of the power supply must be the same, so $z = 3.0$ A.

② (a) 15 J
(b) 2.4 $\Omega$
(c) 1.8 V
(d) 0.9 W

③ 19.6 A

④ (a) 55 $\Omega$
(b) 13.2 $\Omega$

⑤ (i) $R = \dfrac{10 \times 30}{10 + 30} = \dfrac{300}{40} = 7.5\ \Omega$
(ii) $V = 25 \times \dfrac{7.5}{7.5 + 5} = 25 \times \dfrac{7.5}{12.5}$
$= 15$ V
(iii) 22.5 W

⑥ 40 $\Omega$

⑦ (a) 3 W
(b) 1.3 $\Omega$

⑧ $E = 9.0$ V; $r = 0.3\ \Omega$

⑨ $E = 1.65$ V; $r = 2.5\ \Omega$; $I_{max} = 0.66$ A

## Section 2.4

① Horizontal

② 50 Hz

③ $f = 50$ Hz; $T = 20$ ms

④ (a) (i) $\lambda = 0.40$ m;
(ii) $v = 0.5$ m s$^{-1}$;
(iii) $f = 1.25$ Hz
(b) (i) amplitudes are the same;
(ii) phase difference = ¼ cycle

⑤

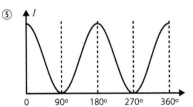

## Section 2.5

① The sound wave diffracts round the door jamb because its wavelength is similar to the size of the doorway. The light waves do not because their wavelengths are much smaller.

② For 100 MHz, $\lambda$ = 3 m so valley width >> $\lambda$.
For 1 MHz, $\lambda$ = 300 m so more diffraction.

③ $S_1R - S_2R = \frac{3}{2}\lambda$

④ The intensity of the fringes follows the diffraction pattern – significant light is diffracted only to small angles.

⑤ Because the unit of $\lambda$ is m; we must have m$^2$ on the top and m on the bottom.

⑥ 630 nm (2 s.f.)

⑦ $2 \times 10^{-6}$ m [2 µm]

⑧ (a) $5.9 \times 10^{-7}$ m
(b) 3

⑨ $6.3 \times 10^{-7}$ m

⑩ 384 m s$^{-1}$

⑪ The even harmonics (2nd, 4th, 6th....) because these would have a node in the middle, which is where the string was struck.

⑫ Because the fringes will be too close together to distinguish clearly.

⑬ (a) If $D$ is too small the fringes will be too close together.
(b) The brightness of the light source: the larger $D$, the fainter the fringes.

⑭ (a) $y = 1.875$ mm ± 3% [1.88 ± 0.06 mm].
(b) 0.63 ± 0.05 µm [Hint: keep percentage uncertainties to 2 s.f. until the final calculation].

⑮ (a) $\lambda_1 = 533$ nm; $\lambda_2 = 530$ nm → best value 532 nm.
(b) The wall would have to be very long; e.g. use a vertical board at an angle (e.g. 45°) to the wall.

⑯ 340 ± 6 m s$^{-1}$

## Section 2.6

① (a) 0.875 m
 (b) 0.860 m
 The pitch will be the same because that is determined by the frequency which is unchanged.

② 4.0 m s⁻¹

③ 32°. It is the same angle.

④ 61.0°

⑤ The critical angle
 $c = \sin^{-1}\dfrac{1}{1.31} = 49.8°$.
 The angle of incidence at the air is 45° so the light ray is refracted out. Angle of refraction = $\sin^{-1}$ (1.31 sin 45°)
 ~ 68°

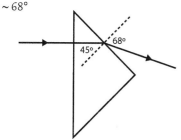

⑥ (a) Speed of light in core =
 $\dfrac{250\ m}{1.28\ \mu s} = 1.953 \times 10^8\ m\ s^{-1}$
 $\therefore n = \dfrac{c}{v} = \dfrac{3.00 \times 10^8\ m\ s^{-1}}{1.953 \times 10^8\ m\ s^{-1}}$
 $= 1.536 = 1.54\ (3\ s.f.)$
 (b) 12.3° [Hint: The path length is proportional to the time].
 (c) At angles to the axis greater than this, the light will be incident on the core-cladding boundary at angles less than the critical angle so TIR will not take place. Without TIR a high percentage of the light will be lost to the core at each incidence with the cladding so the intensity of the light will fall rapidly.

⑦ (a) $n = \dfrac{c}{v} \therefore c = \dfrac{3.00 \times 10^8\ m\ s^{-1}}{1.50}$
 $= 2.00 \times 10^8\ m\ s^{-1}$
 (b) 8.00 μs

## Section 2.7

① $2.0 \times 10^{-18}\ J\ [12.5\ eV]$

② $E_{red}$ (700 nm) = $2.8 \times 10^{-19}\ J$
 $= 1.8\ eV$
 $E_{violet}$ (400 nm) = $5.0 \times 10^{-19}\ J$
 $= 3.1\ eV$

③ Radio; $7.5 \times 10^{26}$ photon s⁻¹

④ 980 nm; infra-red

⑤ $8.75 \times 10^{11}$ electron s⁻¹

⑥ 0.35 eV (!); $5.6 \times 10^{-20}\ J$

⑦ From graph
 $E_{k\ max} = 0.95 \times 10^{-19}\ J = 0.59\ eV$

⑧ (a) Using threshold frequency
 $4.55 \times 10^{14}\ Hz$;
 $f = 3.0 \times 10^{-19}\ J\ [= 1.9\ eV]$
 (b) Using graph points: (7, 1.6) and (0, −3.1):
 $h$ = gradient = $6.7 \times 10^{-34}\ J\ s$

⑨ (a) Intercept on $V_{stop}$ axis = $-\dfrac{\phi}{e}$
 (b) Intercept on $f$-axis = $\dfrac{\phi}{h}$ (this is the same as for the $E_{k\ max}/f$ graph)

⑩ $f_{Thresh} = 4.55 \times 10^{14}\ Hz$.
 $\lambda = 660\ nm$; Red (orange-red)

⑪ $2.9 \times 10^{14}\ Hz$; Infra-red

⑫ $2.18 \times 10^{-18}\ J$

⑬ Transition (b) → 657 nm
 Transition (c) → 121 nm

⑭ Ultraviolet; all other transitions energies between 10.2 eV and 13.6 eV, so they will all be in the u-v.

⑮ The temperature must be high enough for some of the hydrogen atoms in the atmosphere to be in the first excited level.

⑯ (a) $1.47 \times 10^{-21}\ N\ s$
 (b) 37 km s⁻¹

⑰ 206 kW
 Because the leaf could not be 'perfectly' reflecting it would be almost instantly vaporised; even a small fraction of the 206 kW would be enough to do this in short time!

⑱ $3.0 \times 10^7\ m\ s^{-1}$

⑲ 3800 V

⑳ (a) $p = \sqrt{2m_e\ E}$, $E = eV$ and $\lambda = \dfrac{h}{p}$.
 $\therefore \lambda = \dfrac{h}{\sqrt{2m_e eV}} = \dfrac{h}{\sqrt{2m_e e}} \times \dfrac{1}{\sqrt{V}}$, QED
 $b = \dfrac{h}{\sqrt{2m_e e}} = 1.23 \times 10^{-9}\ m\ V^{-0.5}$
 (b) Graph of $r_2$ against $\dfrac{1}{\sqrt{V}}$

㉑ Potentiometer [potential divider]

㉒ red, yellow, green, blue

㉓ Gradient, $\dfrac{hc}{e}$, $\therefore h = \dfrac{gradient \times e}{c}$

## Section 2.8

① $2.87 \times 10^{-19}J$; 1.79 eV

② (a) 1.04 μm
 (b) $2.72 \times 10^{-19}\ J$
 (c) 0.70 (70%)

③ (a) The internodal distance is $\frac{1}{2}\lambda$. So $L = \frac{1}{2}n\lambda$ where $n = 1, 2, 3...$
 $\therefore$ Rearranging $\lambda = \dfrac{2L}{n}$, QED
 (b) For 820 nm, $\dfrac{L}{\lambda} = \dfrac{0.2050\ mm}{820.0\ nm}$
 $= 250$.
 So $L = \frac{1}{2}\ n\lambda$, where $n = 500$
 For 821 nm, $\dfrac{L}{\lambda} = \dfrac{0.2050\ mm}{821.0\ nm}$
 $= 249.69.....$
 So there is no whole number $n$ such that $L = \frac{1}{2}n\lambda$
 (c) The next highest value of $n$ is 501. In this case:
 $\lambda = \dfrac{2 \times 0.2050\ mm}{501}$
 $= 818.4\ nm\ (4\ s.f.)$

## Section 3

① 1°

② 17.3 Ω

③ (a) 30.1 cm [measure to resolution of instrument]
 (b) 7.5 mA [give the unit]
 (c) 6.4 Ω [inconsistent s.f.]

④ Gradient = 0.028 N mm⁻¹
 Intercept = 0.40 N
 Many people do not write units for the gradient and intercept. We recommend that you do as it helps to interpret their

significance, e.g. the unit of the gradient of a velocity-time graph is m s$^{-2}$, so it represents an acceleration.

⑤ 1.92 m s$^{-2}$

⑥ $y_1 \propto \dfrac{1}{x}$ or $y_1 = \dfrac{60.0}{x}$

$y_2$ is linearly related to $x$ – when $x$ increases by 2.0, $y_2$ increases by 1.0

$y_3 \propto x$ or $y = 0.4x$

$y_4 \propto x^2$ or $y_4 = 0.25x^2$

In each case, both answers are correct but the second gives more information.

⑦ The intercept is 0; the gradient is $\dfrac{\lambda}{d}$. To find $\lambda$, measure the gradient and multiply by $d$.

⑧ (a) $\dfrac{4\pi^2}{k}$

(b) e.g. $T$ against $\sqrt{m}$ – gradient $= \dfrac{2\pi}{\sqrt{k}}$, so $k = \dfrac{4\pi^2}{\text{gradient}^2}$

⑨ (a) $\dfrac{1}{V}$ against $\dfrac{1}{R}$

(b) Intercept on $\dfrac{1}{V}$-axis is $\dfrac{1}{E}$ and the gradient is $\dfrac{r}{E}$

So $E = \dfrac{1}{\text{intercept}}$

and $r = \dfrac{\text{gradient}}{\text{intercept}}$

⑩ 221 ± 8 kPa

⑪ 1.4%

⑫ (a) 1.8% or 2%

(b) 15.3 ± 0.3 cm$^3$

⑬ 0.52 ± 0.2 cm$^3$

⑭ 1.64 → 1.90 m s$^{-2}$

$[yT^2 = 1.77 ± 0.13$ m s$^{-2}]$

⑮ $u = 2.5 ± 0.3$ m s$^{-1}$

$a = 1.60 ± 0.05$ m s$^{-2}$

## Answer to AO and skills allocation in the balance question

### AO allocations

(a) (i) Definition of centre of gravity – AO1

(ii) Use of principle of moments and $W = mg$ – AO1

Calculations to give mass – 3 × AO2

(b) Conclusion of the material of the ball – 4 × AO3

### Maths marks allocations

(a) (ii) Calculations to give mass – 3 maths marks

(b) Calculation of density – 3 maths marks

### Practical mark allocations

(a) (ii) Calculation in a practical context – 4 practical marks

(b) Conclusion/evaluation – 4 practical marks.

# Extra questions answers

## Section 1.1

1.  $[\text{RHS}] = \sqrt{[g][d]} = \sqrt{\text{m s}^{-2} \times \text{m}} = \text{m s}^{-1}$

    $[\text{LHS}] = [v] = \text{m s}^{-1} = [\text{LHS}]$

    $\therefore$ Homogeneous QED

2.  (a) For homogeneity $\left[\dfrac{\sigma k}{\rho}\right] = \left[\dfrac{g}{k}\right]$.

    $\therefore [\sigma] = \dfrac{[\rho][g]}{[k^2]} = \dfrac{\text{kg m}^{-3} \times \text{m s}^{-2}}{\text{m}^{-2}} = \text{kg s}^{-2}$

    (b) In part (a), we have ensured that $\dfrac{g}{k}$ and $\dfrac{\sigma k}{\rho}$ have the same units,

    $\therefore$ we only need show that either

    $\sqrt{\dfrac{g}{k}}$ or $\sqrt{\dfrac{\sigma k}{\rho}}$ or have the units of speed

    (the unit of the LHS). Nonetheless, we shall do both:

    $[\text{RHS}] = \left[\sqrt{\dfrac{g}{k}}\right] = \left[\sqrt{\dfrac{g\lambda}{2\pi}}\right] = \sqrt{\text{m s}^{-2}\,\text{m}} = \text{m s}^{-1}$

    ($2\pi$ is dimensionless)

    Or $[\text{RHS}] = \left[\sqrt{\dfrac{\sigma k}{\rho}}\right] = \left[\sqrt{\dfrac{\sigma 2\pi}{\rho\lambda}}\right] = \sqrt{\dfrac{\text{kg s}^{-2}}{\text{kg m}^{-3}\,\text{m}}}$

    $= \sqrt{\dfrac{\text{s}^{-2}}{\text{m}^{-2}}} = \sqrt{\text{m}^2\,\text{s}^{-2}} = \text{m s}^{-1}$

    (c) J = N m. Dividing both sides of this by m²

    $\rightarrow$ J m$^{-2}$ = N m$^{-1}$ QED

    N = kg m s$^{-2}$, $\therefore$ N m$^{-1}$ = kg s$^{-2}$= [$\sigma$] QED

3.  35 km s$^{-1}$

4.  (a) 53 N at E 31.9° S

    (b) $\Delta v$ = 17.3 m s$^{-1}$ at S 17.5° W.

    $\langle a \rangle$ = 1.73 m s$^{-2}$ at S 17.5° W

5.  (a) Resolving vertically and using a bit of common sense (the sled is not going to accelerate into the ground or start to fly).

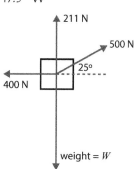

$W = 211 + 500 \sin 25°$

$\therefore W = 211 + 211 = 422$ N

(b) Resultant force = 500 cos 25° − 400 = 53 N horizontally to the right

(c) First calculate the mass of the sled using

$W = mg$

$\therefore m = \dfrac{W}{g} = \dfrac{422}{9.81} = 43$ kg

And using $F = ma$

$\therefore a = \dfrac{F}{m} = \dfrac{53}{43} = 1.23$ m s$^{-2}$

6.  3815 m

7.  Balance point = 34.6 cm.

    [Note: No need to change to kg; keep the forces as $mg$ then cancel by $g$]

8.  (a) Note, in both cases

    $F_A + F_B = 0.180g = 1.77$ N

    (i) $F_A = 1.68$ N; $F_B = 0.09$ N

    (ii) $F_A = 0.59$ N; $F_B = 1.18$ N

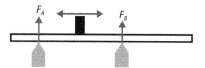

(b) At the 73.8 cm mark

9.  (a) The first thing to say is that the results have been taken by an able student who can read to a debatable accuracy of one tenth of a micrometer division (this level of accuracy is better than expected in A-level practicals).

    Next, there is one debatable point which should be ignored in calculating the mean and uncertainty: 0.185 mm. This point is well outside the range of the other readings.

    Ignoring this reading the diameter is 0.193 ± 0.002 mm

    $\therefore$ The results are consistent with the stated 0.193 mm (except for the one anomalous result).

(b) The precision in $d = \dfrac{0.002}{0.193} = 0.010 \; [= 1.0\%]$

$d$ appears as $d^2$ in calculating $l$, so the precision in $l = 0.020$.

$\therefore \Delta l = \pm 0.02 \times 3815 \text{ m} = \pm 76 \text{ m}$ [using the answer to question 6].

$\therefore l = 3820 \pm 80 \text{ m}$ (taking the uncertainty to 1 s.f.)

## Section 1.2

1. (a) $4.17 \text{ m s}^{-2}$
   (b) $93.8 \text{ m}$
   (c) $9.38 \text{ m s}^{-1}$
   (d) $5\,000 \text{ N}$
2. (a) $5.7 \text{ s} - 7.3 \text{ s}$
   (b) $-16 \text{ m s}^{-1}$
   (c) $a = \dfrac{v - u}{t} = \dfrac{-16 - 10}{4} = -8.5 \text{ m s}^{-2}$
3. (a) $0 \text{ s}, 15 \text{ s}, 25 \text{ s}$
   (b) $13.5 \text{ s}$ (same for both)
   (c) $8.89 \text{ m s}^{-1}$
4. (a) The total distance of decent.
   (b) The acceleration due to gravity.
   (c) C–D large deceleration due to large air resistance when parachute opened

   D–E air resistance decreases until equal to weight – deceleration decreases

   E–F second lower terminal velocity drag = weight again

   F–G sudden deceleration when sky diver hits the ground.
   (d) $425 \text{ N}$
5. (a) $17.5 \text{ m s}^{-1}, 17.5 \text{ m s}^{-1}$
   (b) $15.5 \text{ m}$
   (c) $31.1 \text{ m}$
6. (a) $21.2 \text{ m s}^{-1}$ at an angle of $34.4°$ below the horizontal.
   (b) $8.25 \text{ m}$

## Section 1.3

1. (a)
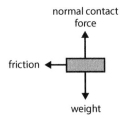

normal contact force

friction

weight

   (b) **weight** – gravitational force acting on the puck due to the Earth,

   N3 partner – gravitational force acting on the Earth due to the puck

   **normal contact force** – upwards force on the puck due to the ice

   N3 partner – contact force pushing down on the ice due to the puck

   **friction** – frictional force to the left on the puck due to its motion on the ice

   N3 partner – frictional force to the right on the ice due to the motion of the puck

2. (a) $4.44 \text{ m s}^{-1}$
   (b) $11.1\%$
3. Velocity of truck after collision $= 3.75 \text{ m s}^{-1}$

   final total KE $= 531 \text{ kJ}$ whereas the initial KE $= 500 \text{ kJ}$, so this is not possible without energy input.
4. (a) $305 \text{ N}$
   (b) (i) $476 \text{ N}$
       (ii) $64.2 \text{ m s}^{-1}$
5. (a) There is a change in the direction of motion of the water and hence its velocity changes. Momentum is mass multiplied by velocity hence the momentum also changes.

   The vector $v - u$ gives the change in velocity
   (b) According to N2, the water experiences a force in the direction of the change in momentum as shown. Using N3, there must be an equal and opposite force exerted by the water on the hose, i.e. its direction is opposite to that shown in the diagram.
   (c) $12.5 \text{ N}$

## Section 1.4

1. (a) $31.1 \text{ m s}^{-1}$

   (b) Kinetic energy $\rightarrow$ Thermal energy (internal energy of brakes)

   (c) $5160 \text{ N}$

2. (a) $442 \text{ J}$

   (b) $10 \text{ m s}^{-1}$

   (c) $29.7 \text{ m s}^{-1}$

3. (a) Initial KE of lorry becomes thermal (internal) energy of the brake disks and pads. This is gradually dissipated and although not destroyed, cannot be recovered as useful energy.

   (b) There is no useful output energy (unless the lorry is a hybrid and has regenerative braking). Zero useful energy substituted into

   $$\text{efficiency} = \frac{\text{useful energy transfer}}{\text{total energy input}} \times 100\%$$

   gives an efficiency of 0%.

4. (a) $8830 \text{ J}$

   (b) $36.8\%$

   (c) Total work done = $400 \text{ N} \times 60 \text{ m} = 24\,000 \text{ J}$

   Work done in lifting bricks = $8830 \text{ J}$

   Work done in lifting pulley and pallet
   $$= (5 + 2) \times g \times 15 = 1030 \text{ J}$$

   Other wasted energy
   $$= 24\,000 - 8830 - 1030 = 14\,140 \text{ J}$$

   (from conservation of energy)

   This other wasted energy would manifest itself as thermal (internal) energy of the pulley and rope due to friction. Air resistance in this case would be negligible because of the low speeds. However, there could also be a small but significant amount of kinetic energy not vertically but horizontally due to the pendulum motion of the bricks, pallet and pulley.

5. $84.9 \text{ m s}^{-1}$

## Section 1.5

1. $\text{kg s}^{-2}$

2. $0.584 \text{ N cm}^{-1}$ ($58.4 \text{ N m}^{-1}$)

3. $93.8 \text{ J}$

4. (a) EPE = $\frac{1}{2}Fx$; the force is the same for both. However, the thick rod of csa $2A$ will extend half as much (since $\Delta l = \dfrac{Fl_0}{AE}$) meaning that it will store half the energy of the thinner rod.

   (b) EPE = $\frac{1}{2}Fx$; this time the force is twice as much in the thinner rod. Also, the extension in the thinner rod will be 4 times greater ($\Delta l = \dfrac{2Tl_0}{AE}$ compared with $\dfrac{Tl_0}{2AE}$). Hence, the thinner rod will store 8 times more energy.

5. (a) (i) $560 \text{ N}$

      (ii) $0.577 \text{ m}$

      (iii) $162 \text{ J}$

   (b) The shape of the graph cannot be known perfectly but we can make a guess at the mean stress.

   Mean stress $\approx 120 \text{ MPa}$

   Work done = area under graph $\times$ volume of wire
   $$= 448 \text{ kJ}$$

from a(i) 90% of this strain gives an extension of 0.577 m i.e. strain = 0.000641

6. Gradient of graph of $L/g$ against $\Delta x$ / mm = 202

$$\therefore \frac{F}{\Delta l} = 202 \times \frac{1 \times 10^{-3} \times 9.81}{1 \times 10^{-3}} = 1982 \text{ N m}^{-1}$$

$$\therefore E = \frac{F}{\Delta l} \times \frac{l_0}{A} = 1982 \text{ N m}^{-1} \times \frac{3.43 \text{ m}}{\pi (0.136 \times 10^{-3} \text{ m})^2}$$

$$= 106 \text{ GPa}$$

## Section 1.6

1. Mainly infra-red with some visible. [Peak wavelength $\sim$ 970 nm]

2. (a) (i) [Extreme] ultraviolet. The spectral intensity decreases with wavelength in the visible region of the e-m spectrum [not shown], so the blue radiation dominates.

   (ii) $T = \frac{W}{\lambda_n} = \frac{2.90 \times 10^{-3} \text{ m k}}{55 \times 10^{-9} \text{ m}} = 53\,000$ K

   $\sim 5 \times 10^4$ K QED

   (b) (i) $2.11 \times 10^{33}$ W

   (ii) Luminosity = $5.5 \times 10^6\ L_{Sun}$

   (iii) $3.9 \times 10^{10}$ m

3. (a) Spectral radiance is the power per unit area per unit wavelength interval.

   Hence unit = W m$^{-3}$. [Note: If wavelength is expressed in nm the unit of spectral radiance is often given as W m$^{-2}$ nm$^{-1}$]

   (b) This is the total power of the radiation per unit area.

4. (a) 12 pm; X-ray

   (b) $3.4 \times 10^{31}$ W

   (c) Temperature is multiplied by $\frac{1}{\sqrt[4]{2}} \rightarrow$ Temperature reduced to $2.1 \times 10^7$ K

5. (a) The temperature in the accretion disk of a supermassive black hole can rise to such high temperatures that the thermal radiation emitted is in the $\gamma$ ray region. The cosmological red shift, when observing from large distances, can cause the radiation to be red-shifted into the X-ray and u-v regions.

(b) The colour of a star depends upon the wavelength of peak emission. A blue/white star is one with $\lambda_{max}$ at the blue end of the visible spectrum and is therefore a hotter star than a red/orange one. Thus a blue-white main sequence star is heavier than a red/orange one.

(c) (i) Visible for reflections of the sun's radiation; infra-red for thermal emissions

   (ii) $\gamma$-ray

   (iii) Microwave / far infra-red [so-called sub-millimetre or terahertz radiation]

   (iv) Infra-red [because nebula is more transparent to the infra-red from the hot dense regions]

   (v) Microwave

   (vi) X-ray [and extreme u-v]

## Section 1.7

1. $U$: 4 on both sides; $D$: 2 on both sides

2. Only quarks (and hadrons) feel the strong nuclear force so the neutrino (lepton) does not feel it. The neutrino is not charged; neither is it composed of charged particles (like the neutron – udd) and so cannot feel the e-m force.

3. $\quad \pi^+ \rightarrow e^+ + \nu_e$

   $Q \quad 1 \rightarrow \quad 1 + 0$

   $B \quad 0 \rightarrow \quad 0 + 0$

   $L \quad 0 \rightarrow \quad -1 + 1$

   There is a change of quark flavour ($u\bar{d} \rightarrow$ no quarks) and there is an electron neutrino involved. These are both tell-tale signs of the involvement of the weak nuclear force.

4. $\pi^- \rightarrow e^- + \bar{\nu}_e$

5. $\rho^+ \rightarrow \pi^+ + \pi^0$.

   From baryon number conservation, the hadron must have a baryon number of zero so it must be a meson. From conservation of charge, the hadron also has to be neutral. Hence $\pi^0$ is the only possibility left.

6. $^{7}_{4}\text{Be} + e^{-} \rightarrow {}^{7}_{3}\text{Li} + Y$

   In $^{7}_{4}\text{Be}$ one proton decays to a neutron to give $^{7}_{3}\text{Li}$. At the quark level this means that a u quark becomes a d quark.

   $$u + e^{-} \rightarrow d + Y$$

   | | | |
   |---|---|---|
   | $Q$: | $\tfrac{2}{3} - 1 \rightarrow -\tfrac{1}{3} + Q_Y$ | $Q_Y = 0$ |
   | $B$: | $\tfrac{1}{3} + 0 \rightarrow \tfrac{1}{3} + B_Y$ | $B_Y = 0$ |
   | $L$: | $0 + 1 \rightarrow 0 + L_Y$ | $L_Y = 1$ |

   The only particle we know of with baryon number of zero, charge of zero and lepton number +1 is an electron neutrino. Hence Y is an electron neutrino. (This process is known as electron capture.)

7. Strong nuclear force (rearrangement of neutrons and protons, all particles are hadrons, no change in quark flavour).

8. $^{37}_{17}\text{Cl} + \nu_e \rightarrow {}^{37}_{18}\text{Ar} + e^{-}$

9. The weak interaction is responsible. This means that (a) the neutrinos have to approach the neutrons very closely to interact (they have to hit them) and (b) even then the interaction is very unlikely to happen.

10. (i) Not possible, violates charge conservation.

    (ii) Not possible, violates baryon number conservation.

    (iii) Not possible, violates baryon number conservation.

    (iv) Possible, satisfies all 3 conservation laws, strong force.

    (v) Possible, satisfies all 3 conservation laws (electron capture, weak force).

    (vi) Not possible, violates charge conservation, baryon number conservation and lepton number conservation.

    (vii) Possible, satisfies all 3 conservation laws, strong force.

## Section 2.1

1. Electrons have moved from the hair to the comb. The comb has a charge of –25 pC while the hair will have a charge of +25 pC.

   This corresponds to the movement of
   $$\frac{25 \times 10^{-12}}{1.6 \times 10^{-19}} = 1.56 \times 10^{8} \text{ electrons}$$

2. (a) Current is the rate of flow of charge and charge is the (mean) current multiplied by the time. Hence, A s is the base unit of charge but A h is also a (larger) unit of charge.

   (b) $500 \times 10^{-3}$ A $\times 60 \times 60$ s $= 1800$ C

3. (a) 0 (+92$e$ for the protons and –92$e$ for the electrons)

   (b) +3$e$ = $4.8 \times 10^{-19}$ C

   (c) 578 kC

4. Approximately as shown

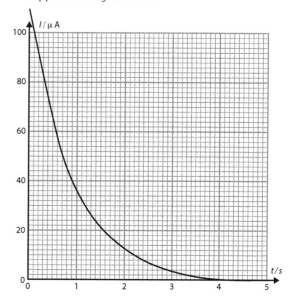

5. (a) A has 12× as many electrons (4 × area, 3 × length)

   (b) Same (they are both the same material)

   (c) Drift velocity in B is 8 × as much (one quarter the area but twice the current)

## Section 2.2

1. (a) 30 W
   (b) 65.2 mA
   (c) 111 Ω
   (d) 14.0 mW
   (e) 124 V

2. (a) 58.8 Ω
   (b) About 13 min assuming the resistance of the element is the same. (The resistance will be actually be slightly smaller because of the lower power and therefore the element temperature.)

3. $1.44 \times 10^{-6}$ Ω m

4. (a) Same resistance.
   (b) Drift velocity is 4 times as high in B.

5. 226 Ω assuming a constant rise of 0.367 Ω per °C (Not a very good assumption)

## Section 2.3

1. 4.7 Ω, 6.8 Ω and 8.2 Ω using single resistors

   2 resistors in series: 11.5 Ω, 15.0 Ω and 12.9 Ω

   3 resistors in series 19.7 Ω

   2 resistors in parallel 2.8 Ω, 3.7 Ω and 3.0 Ω

   3 resistors in parallel 2.1 Ω

   1 resistor in series with 2 parallel 8.4 Ω, 9.8 Ω and 11.0 Ω

   1 resistor in parallel with 2 in series 3.6 Ω, 4.5 Ω and 4.8 Ω

2. 207 Ω

3. 32.0 V, 0.72 A

4. 

[Graph: $V_{OUT}$/V (y-axis, 0 to 6.0) against Temperature / °C (x-axis, −20 to 40)]

5. It is used for 4 hours a day.

6. As light falls on $R_L$, its resistance will decrease. Its fraction of the resistance of the potential divider will decrease, hence the pd across it will decrease.

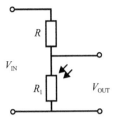

## Section 2.4

1. (a) Because the wave could have travelled more than one wavelength between the two snapshots. In that case the speed would be greater than that calculated.
   (b) Next highest: Wave has travelled $9 \times 0.05$ m $\rightarrow v = 4.5$ m s$^{-1}$; $f = 11.25$ Hz

   Next highest: Wave has travelled $17 \times 0.05$ m $\rightarrow v = 8.5$ m s$^{-1}$; $f = 21.25$ Hz

2. (a) 10 m s$^{-1}$
   (b) Wave would be ¼ of a cycle further to the right (i.e. peaks at 0.25 m, 0.45 m.)
   (c) Because the oscillations of the particles in the wave are verticle which is at right angles to the direction of motion of the wave.

3. (i) 0.7 s
   (ii) C and D are in phase with A: they are the same distance to the right of the nearest wave crest.

   B is out of phase – it is to the left of the nearest wave crest.

4. (a) (i) $\lambda = 42$ m
   (ii) $f = 0.20$ s
   (iii) $v = 8.4$ m s$^{-1}$
   (b) ~ 10 m; ~ 52 m; ~ 93 m
   (c) From the steepest gradient of the displacement–time graph ~ 2 m s$^{-1}$

## Section 2.5

1. (a) 670 nm

   (b) (i) The light from the laser spreads out after passing through the slit.

   (ii) Without diffraction the beams from the two slits would not overlap. Only if a point on the screen receives light from both slits can interference occur at that point.

   (c) The phase of the light arriving from slits 1 and 2 ($S_1$ and $S_2$) to a point, P, depends on the path lengths $S_1P$ and $S_2P$. Assuming the light in the slits is in phase, then if $|S_1P - S_2P| = (n + \frac{1}{2})\lambda$, where $n = 0, 1, 2...$, the light from $S_1$ and $S_2$ will arrive at P in antiphase and so will interfere destructively, producing a dark fringe. The term $S_1P - S_2P$ is called the path difference.

   (d) The fringe separation would increase by a factor of $\frac{7.5}{1.5}$ = 5 to 10 mm. The brightness of the fringes would be $\frac{1}{5^2}$ = 0.04 times as much.

2. (a) $n\lambda = d\sin\theta$, $\therefore d = \frac{n\lambda}{\sin\theta} = \frac{2 \times 532 \text{ nm}}{\sin 28.9°}$

   $= 2200$ nm

   $\sim 2 \times 10^{-6}$ m QED

   (b) (i) 632 nm

   (ii) 7 [up to 3rd order]

3. (a) $\lambda = 1.60$ m

   (b) Time = 5.0 ms

   (c) (i) Amplitude of B $\sim 0.5 \times$ Amplitude of A

   (ii) Phases of A and B the same

   (d) (i) Amplitude of B = amplitude of C

   (ii) B and C in antiphase

   (e) 16.7 Hz

4. (a) 489 nm

   (b) 487 nm [a pretty good approximation]

## Section 2.6

1. 1.50

2. (a) 80.1°

   (b) A light ray travelling in sea water is incident upon a block of ice with an angle of incidence greater than 80.1°

3. (a) 6.6°

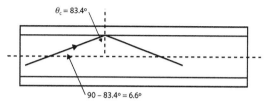

$\theta_c = 83.4°$

$90 - 83.4° = 6.6°$

   (b) The difference in propagation time for light travelling along the axis and light hitting the core-cladding boundary at the critical angle (the extreme conditions) is lower, for lower propagation angles to the axis. Thus signals can travel further before neighbouring bits of data start interfering with one another.

4. (a) 72.7°

   (b) After $n$ reflections,
   Power remaining $= P \times (0.1)^n = \frac{P}{10^n}$

   $\therefore$ To calculate $n$ we need the solution of:
   $\frac{P}{10^n} = \frac{P}{10^6}$
   $\therefore n = 6$.

   $\therefore$ There are 5 reflections subsequent to P.

   $\therefore$ Distance along rod = $5 \times 12$ mm
   $= 60$ mm.

   (c) 65.1°

5. The normal direction, i.e. the one along the direction of the change of wave speed is vertical. Consider the angle, $\theta$, that the sound 'rays' (i.e. the direction of propagation of the waves) makes with the vertical:

$\sin \theta \propto v$ where $v$ is the velocity of propagation.

∴ $\theta$ decreases with height and so the sound travels closer to the vertical.

[This explains the *hush* on a cold still frosty morning. Sounds that people and birds make are refracted over your head.]

6. For the light ray in Fig. 2.6.4:
$n_a \sin \theta_a = n_g \sin \theta_g$ and $n_g \sin \theta_g = n_w \sin q_w$.
∴ We can eliminate $n_g \sin \theta_g$ to give
$n_a \sin \theta_a = n_w \sin \theta_w$,
which is the same equation as in Quickfire 3 and so the presence or absence of the glass is irrelevant for the direction of the light ray inside the fish tank. This is only true because the glass is parallel sided so the angle of refraction $\theta_g$ going into the glass has the same value as the angle of incidence $\theta_g$ going out of the glass.

7. The partial reflections and transmission become fainter with each successive incidence.

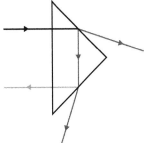

## Section 2.7

1. (a) $\phi$
   (b) $hf$

2. (a) When emission occurs the maximum KE = photon energy – work function;
   Photon energy = 2.90 eV.
   ∴ $E_{k\,max}$ = 2.90 eV – 2.10 eV = 0.80 eV
   (b) Photon energy = 2.36 eV
   ∴ $E_{k\,max}$ = 2.36 eV – 2.10 eV = 0.26 eV
   (c) Photons do not combine so the maximum photon energy determines the maximum electron energy. So $E_{k\,max}$ = 0.80 eV
   (d) Photon energy = 1.86 eV. This is less than the work function (2.10 eV) so no electrons are emitted.

3. (a) 13.6 eV
   (b) (i) If photons of the correct energy are incident on a hydrogen atom in the ground state, the –2.18 × 10⁻¹⁸ J level, they can be absorbed. Electrons are promoted to the one of the higher energy levels. These atoms subsequently re-emit the radiation in random directions thus the radiation in the forward direction is reduced.
   (ii) The photon energy is given by
   - $\Delta E$ = (2.18 – 0.54) × 10⁻¹⁸ J
     = 1.64 × 10⁻¹⁸ J
   or
   - $\Delta E$ = (2.18 – 0.24) × 10⁻¹⁸ J
     = 1.94 × 10⁻¹⁸ J
   The wavelengths of these photons are 143 nm and 121 nm respectively.
   (iii) For the first and second excited states, $\Delta E$ = 0.30 × 10⁻¹⁸ J = 1.88 eV. This is the photon energy for light of wavelength 663 nm, which is in the red end of the visible spectrum.

4. (a) $E$ = 3.16 × 10⁻¹⁹ J = 1.97 eV;
   $p$ = 1.05 × 10⁻²⁷ N s
   (b) Assume a monochromatic beam with of wavelength $\lambda$.
   Photon energy $= \dfrac{hc}{\lambda}$;
   photon momentum, $p = \dfrac{h}{\lambda} = \dfrac{E}{c}$
   Beam power = $P$, so the number of photons per second $= \dfrac{P}{E}$
   ∴ The momentum of the incident photons per second $= \dfrac{P}{E} \times \dfrac{E}{c} = \dfrac{P}{c}$
   ∴ The momentum of the reflected photons per second $= -\dfrac{P}{c}$
   ∴ The momentum change per second of the photons $= -2\dfrac{P}{c}$
   ∴ The force exerted by the surface on the light beam $= -2\dfrac{P}{c}$
   ∴ By N3, the force exerted by the light beam on the surface $= 2\dfrac{P}{c}$

5. (a) 530 keV

(b) The kinetic energy, $E = \dfrac{p^2}{2m}$. Hence, for the same energy, the larger the mass the larger the momentum. So the electron has a larger momentum than the proton.

But $p = \dfrac{h}{\lambda}$ so the electron has a smaller wavelength than the proton.

## Section 2.8

1.

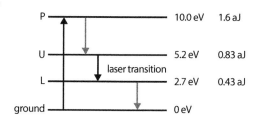

2. (a) (i) Level O is full; levels U and P are almost empty.

(ii) Absorption.

The photon is absorbed and its energy used to promote an electron from level O to U.

(b) (i) The population of level U is greater than that of level O. It is achieved by pumping electrons from level O to the short-lived state P. These drop down rapidly into the metastable state U as shown. If more than half the electrons from O are promoted then the number in level U can exceed that in O.

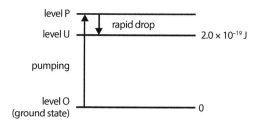

(ii) When an electron drops down to level O it gives rise to a photon of energy $2.10 \times 10^{-19}$ J. This photon interacts with a second atom in state U to cause stimulated emission in which a second photon of the same frequency, phase and direction as the first is produced. Thus there are now two photons instead of one. This process continues with the number of photons increasing exponentially, i.e. the light is amplified.

(iii) 947 nm.

# Practice question answers

① **Model answer**: The resultant force is zero and the resultant moment about any point is zero.

**Alternative answer**: The [vector] sum of the forces is zero and the clockwise moment about any point is equal to the anticlockwise moment about the same point.

**Commentary**: Candidates often forget about the moment. If the resultant moment isn't zero, the body will start rotating. It is important that all moments are taken about the same point.

② **Model answer**: A black body is one which absorbs all electromagnetic radiation which is incident upon it.

**Commentary**: This is the easiest definition. Often candidates mistakenly think that they need to give a definition in terms of the *emission* of radiation, perhaps because a star is clearly radiating. It is possible to give a definition in terms of emission – but it is less satisfactory.

**Alternative answer**: A black body is one which emits the maximum possible amount of radiation at every wavelength.

**Unsatisfactory answer**: A black body is one which absorbs all wavelengths [or frequencies] of electromagnetic radiation which are incident upon it.

**Commentary**: This answer is not credited because it is not clear that all e-m radiation is absorbed – it could be that 50% of the radiation is absorbed at all wavelengths.

③ **Model answer**: A transverse wave is one in which the particle oscillations are at right angles to the direction of propagation of the wave.

Example: Electromagnetic waves / seismic S-waves

A longitudinal wave is one in which the particle oscillations are in line with / parallel to the direction of propagation of the wave.

Example: Sound waves / seismic P-waves

**Commentary**: The answer needs to make clear which two directions are at right angles, or in line as appropriate. The definition given is not entirely general – it covers the case of a wave which consists of oscillations of a medium. Electromagnetic waves consist of coupled oscillations of electric and magnetic fields rather than particles but the examiners will not expect reference to this.

④ **Model answer**: This means that 5.0 J of energy is transferred per unit of charge [or for every coulomb of charge] which passes through the component.

**Less acceptable answer**: 5.0 J of energy is converted into heat when 1 C of charge passes through . . . (?)

**Commentary**: Why is the second answer not as good as the first?

- 'Heat' is itself an energy transfer, so the expression isn't good, but more importantly (for AS) the energy transfer could be to mechanical work, e.g. in a motor.

- The 2nd 'definition' doesn't allow for quantities of charge other than 1 C, while the first one clearly implies that when 2 C of charge passes, 10 J of energy is transferred, etc. This fits in with the unit equivalence $V = J\,C^{-1}$.

⑤ **Model answer**: For a metallic conductor at a constant temperature the current is [directly] proportional to the potential difference.

**Commentary**: There are two marks – one mark is for the proportionality and the other for the conditions under which it applies.

**Unsatisfactory answer**: $V = I \times R$ [or expressed in any other way].

**Commentary**: Why is this not accepted? The equation is just the definition of electrical resistance ['to find the resistance, divide the pd by the current']. It doesn't even state that $V$ and $I$ are proportional – it would need to add that R is a constant. The equation $V = I \times R$ can be used to calculate the resistance of a component for which $V$ and $I$ are not proportional, e.g. a light bulb filament or a diode, although there will be a different answer for different values of $V$ and $I$.

⑥ **Model answer**: [This applies to superconductors.] As a conductor is cooled, the transition temperature is the temperature at which the resistance suddenly drops to zero.

**Less satisfactory answer**: This is the temperature, below which a material is superconducting (?).

**Unsatisfactory answer**: The temperature at which the resistance of a superconductor is zero.

**Commentary**: The answer needs to make clear:

- There is a sudden change in resistance – the second statement is weak here.

- The zero resistance state is for all temperatures [or at least a range of temperatures] below the transition temperature – the third statement suggests it is only zero at that temperature.

⑦ **Model answer**: When two [or more] waves pass through [or exist at] a point, the [total] displacement of the medium at that point is equal to the [vector] sum of the displacements due to the individual waves [at that point].

**Commentary**: There are two parts to the answer. The first one describes the situation, i.e. it deals with two waves. The second describes how they are combined. The words in square brackets are good to have but are not strictly necessary: if the principle works for two waves, it will hold for three or more; the displacement of the medium must be the total displacement; there is only one way to add displacements, i.e. as vectors!

⑧ **Model answer**: A hadron is a particle which is composed of quarks. ✓

A baryon consists of 3 quarks, e.g. a proton [neutron, delta, $\Delta$++, etc.] ✓

A meson consists of a quark and an antiquark, e.g. a pion [$\pi^+$, $\pi^0$, etc.]. ✓

**Commentary**: It is unlikely that an examiner would ask the question in quite this form. Asking for an example of, say, a baryon, means that the examiner must be prepared to accept any out of a vast zoo of baryons, $\Omega^-$, $\Sigma^+$, etc. An alternative definition of the hadron is a composite particle which experiences the strong interaction.

⑨ **Model answer**: Multimode dispersion.

**Commentary**: If you cannot remember the name, give a description in the hope that the mark scheme will allow for it, e.g. in fibres with a wide core, some light rays pass down the middle and others have a zig-zag path and so take different times to reach the other end.

⑩ **Model answer**: 'EMF of 10.0 V' means that 10.0 J of energy is transferred from an external source ✓ per unit charge which passes through the supply ✓.

'PD across its terminals is 9.0 V' means that 9.0 J of energy is transferred to the external circuit ✓ per unit charge which passes through the supply.

**Commentary**: The unit of both EMF and pd is the volt, which is defined as the joule per coulomb, so the explanation must involve both the energy transfer and the expression 'per unit charge' or 'per coulomb'. The context of the energy transfer needs to be given to distinguish the two quantities: EMF refers to energy into the electrical system, e.g. from chemical energy in the cell, or mechanical work done on a dynamo; pd refers to the energy provided by the power supply to the external circuit [this is often more accurately described as the 'electrical work done on the external circuit'] per unit charge. In most cases, the examiner will look for the expression 'per unit charge' used correctly at least once in the answer.

## Experimental description questions

⑪ **Model answer**:

(a) The resistance of a length of conductor is

given by: $R = \dfrac{\rho l}{A}$

So $\rho = \dfrac{RA}{l}$

so $[\rho] = \dfrac{\Omega\ m^2}{m} = \Omega\ m$

(b) Apparatus required: Ohm-meter, metre rule, micrometer.

**Method**: Measure the resistance of a length of wire using the ohm-meter. First check the resistance of the ohm-meter leads by touching them together – subtract this from the reading for the resistance of the wire.

Measure the length, $l$, of the wire using the metre rule.

Measure the diameter, $d$, of the wire, at several points, using the micrometer, determine the mean value and calculate the cross-sectional area using the equation:

$$A = \frac{\pi d^2}{4}$$

Finally, calculate the resistivity using the equation

$$\rho = \frac{RA}{l}$$

**Commentary**:

(a) The equation given in the data sheet needs to be manipulated so that the resistivity is the subject. The unit equation is written, i.e. the equation with the units of the quantities instead of the quantities themselves. You need to do enough, so that the examiner believes you know what you are doing!

(b) The apparatus used is stated. There is no need for a diagram here. The necessary measurements are stated, including avoiding obvious uncertainties, such as the zero-error of the ohm-meter.

How you use the measurements to give the final answer is stated – in this case, how $A$ and, finally, $\rho$ are determined.

⑫ **Model answer**: The wire is placed into a beaker of water as shown and the beaker heated gradually using a Bunsen burner [on tripod and gauze]. The resistance and temperature readings are taken by removing the Bunsen and stirring [thermometer] to allow the temperature to become uniform and then the readings on the thermometer and resistance meter noted. A series of readings up to about 100°C are obtained. For greater accuracy the resistance of the leads to the coil is measured and subtracted to obtain the resistance of the coil inside the liquid.

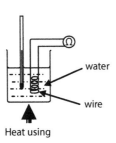

water

wire

Heat using

**Commentary**: The set-up is clear – this helped by a diagram. There is no need to label standard pieces of apparatus [beaker, thermometer, ohm-meter]. The procedure is clear enough for another student to follow it.

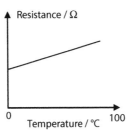

Resistance / Ω

Temperature / °C

0

100

The graph is shown as straight with a non-zero intercept on the resistance axis. In reality it is not exactly straight, but it is very nearly straight over this limited temperature range.

⑬ **Model answer**:

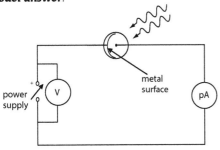

metal surface

power supply

V

pA

[Note additions to diagram]. The pd of the power supply is increased from zero and the current monitored. As the pd increases, the current drops and the minimum pd, $V_S$, for which the current is zero is noted. The maximum kinetic energy is $eV_S$.

**Commentary**: The three modifications to the circuit – polarity of the supply, variable power supply and voltmeter – are shown. The experimental technique, including the necessary statement, 'starting from zero', is given; the necessary measurement stated and the method of calculation given.

(b) (i) Determine the gradient of the graph – this is the Planck constant.

(ii) Extend the graph until it reaches the vertical axis. The intercept is the negative of the value of the work function, i.e. intercept = $-\phi$.

**Commentary**: Note that the command word is 'state' – there is no need for any explanations.

(b) (ii) There is arguably no need for the first sentence but it does make the answer clear – the line could be added to the graph on the question paper. One of the marks is for the 'negative' in the answer. There is an alternative answer, which is 'note the intercept, $f_{threshold}$, on the frequency axis. The work function is $h \times f_{threshold}$.'

## Questions to test understanding

⑭ **Model answer**:

Units of r.h.s. $= \sqrt{\dfrac{N}{kg\ m^{-1}}}$

$= \sqrt{\dfrac{kg\ m\ s^{-2}}{kg\ m^{-1}}}$

$= \sqrt{m^2\ s^{-2}}$

$= m\,s^{-1} =$ units of l.h.s.

So the equation **is** homogeneous, i.e. it is possible in terms of units.

**Commentary**: The working started with the complicated [right-hand] side, expressed the quantities on that side in terms of their units, and simplified the resulting expression. The resulting expression was shown to be the same as the units of the left-hand side. The use of a single cascade of equations, each starting on a new line, with the equals signs aligned, considerably aids communication.

⑮ **Model answer**:

(a) 19 s – 20 s: The acceleration is zero because the upwards air resistance is exactly equal and opposite to the gravitational force on the skydiver, making the resultant force zero.

20 s – 21 s: The parachute opens, so the air resistance very rapidly increases. There is a large resultant upwards force on the parachutist, producing a large deceleration (negative acceleration).

21 s – 25 s: As the parachutist slows down, the air resistance decreases so the resultant upwards force decreases and the negative acceleration decreases in magnitude.

(b) The total area between the time axis and the line gives the velocity. This velocity must be positive (downwards) because the parachutist is moving downwards. QED.

**Commentary**: In fact, the area for the first 20 s will give the first terminal velocity (around 45 m s$^{-1}$ if you want to work it out). The area for the next 17 s is around −40 m s$^{-1}$. This means that the final terminal velocity will be around 5 m s$^{-1}$.

⑯ **Model answer**:
Resistance of parallel combination $= \dfrac{10 \times 15}{10 + 15} = 6\ \Omega$

Using the potential divider formula:
pd across 12 $\Omega = \dfrac{12}{12 + 6} \times 9.0\ \text{V} = 6.0\ \text{V}$

Alternative to the second stage:
Total resistance $= 12 + 6 = 18\ \Omega$
∴ Total current [= current through the 12 $\Omega$ resistor] $= \dfrac{V}{R} = \dfrac{9.0}{18} = 0.5\ \text{A}$
∴ pd across 12 V resistor $= IR = 0.5 \times 12 = 6\ \text{V}$.

**Commentary**: Questions like this require repeated use of simple equations, e.g. $V = IR$, to different aspects of the circuit. It is very important to make it clear which components are referred to in your working, hence you need to say, for example, 'resistance of the parallel combination'.

⑰ **Model answer**:
(a) Total resistance in the circuit $= r + R$
$= 0.5 + 12.5 = 13.0\ \Omega$.
∴ Current $= \dfrac{\text{emf}}{\text{total resistance}} = \dfrac{1.5}{13} = 0.115\ \text{A}$

(b) Current $= \dfrac{\text{emf}}{\text{total resistance}} = \dfrac{1.5n}{12.5 + 0.5n}$

If current $\geq 0.6$ A; $\dfrac{1.5n}{12.5 + 0.5n} \geq 0.6$

∴ $1.5n \geq 0.6\,(12.5 + 0.5n)$
[Expanding the bracket] ∴ $1.5n \geq 7.5 + 0.3n$
[Rearranging] ∴ $1.5n - 0.3n \geq 7.5$
i.e. $1.2n \geq 7.5$ so $n \geq 7.5/1.2 = 6.25$.
So the minimum number of cells is 7 [it must be a whole number]

**Commentary**: The idea that EMF is the 'total voltage' in the circuit is very useful in this type of question.

(b) This is an ideal answer but trial and error is also quite acceptable. If you don't like using the inequality sign [$\geq$] an alternative is to use the equality sign [=], arrive at the answer 6.25 and then make the statement that the smallest whole number which will do is the next one up, i.e. 7.

⑱ **Model answer**:
(a) Rearranging the equation: $k = \dfrac{F}{v^2}$.
So unit of $k$ = N (m s$^{-1}$)$^{-2}$ = N m$^{-2}$ s$^2$.

(b) (i) Resultant force at 20 m s$^{-1}$
$= ma$
$= 75\ \text{kg} \times 8.2\ \text{m s}^{-2}$
$= 615\ \text{N}$
∴ Air resistance at 20 m s$^{-1}$
$= mg - 615\ \text{N}$
$= 75\ \text{kg} \times 9.81\ \text{N kg}^{-1} - 562.5\ \text{N}$
$= 121\ \text{N}$
∴ $k = \dfrac{F}{v^2} = \dfrac{121}{20^2} = 0.302\ \text{N m}^{-2}\ \text{s}^2$
At terminal velocity,
air resistance $= mg = 735.75\ \text{N}$
∴ Terminal velocity,
$v = \sqrt{\dfrac{F}{k}} = \sqrt{\dfrac{735.75}{0.302}}$
$= 49.4\ \text{m s}^{-1} \sim 50\ \text{m s}^{-1}$ QED

(ii) Using the middle value of velocity [40 m s$^{-1}$] to estimate the resultant force:
Estimated resultant force
$= 75 \times 9.81 - 0.30 \times 40^2 = 256\ \text{N}$
Acceleration $= \dfrac{F}{m} = \dfrac{256}{75} = 3.41\ \text{m s}^{-2}$.
Time taken $= \dfrac{\text{change in velocity}}{\text{acceleration}} = \dfrac{10}{3.41} \sim 3\ \text{s}$

**Commentary**:
(a) 'Show that' means that the working must be clear. In this case, the initial rearranging of the equation followed by the algebra – the examiner must believe you knew what you were doing!

(b) (i) The physical principle of the answer is clear and well expressed. In an exam, full marks would be given for just the final answer [if it were correct!]

(ii) This is a 'show that', so there would be no marks for just stating the final answer – the reasoning must be clear.
The word 'estimate' does not mean 'guess' – you need to do a calculation with a simplifying assumption. In this case the simplification replaces the varying drag force with a constant typical one from the middle of the velocity range.

⑲ **Model answer**:

In this circuit, the total input voltage, $V_{IN}$ is given by:

$V_{IN} = IR_{Total} = I(R + X)$

$V_{OUT} = IR$ – assuming that the current in the two resistors is the same, i.e. no current is taken by $V_{OUT}$.

Dividing: $\dfrac{V_{OUT}}{V_{IN}} = \dfrac{IR}{I(R + X)}$

Cancelling by $I$ on the r.h.s.: $\dfrac{V_{OUT}}{V_{IN}} = \dfrac{R}{R + X}$ QED

**Commentary**: The working clearly uses $V = IR$ twice and, importantly, gives the context for each step – 1st for the two resistors in series and 2nd for resistor $R$ alone. The assumption is clearly given and the cancellation by $I$ explicitly stated.

⑳ **Model answer**:

(a) The current through $L_1$ splits up equally to pass through $L_2$ and $L_3$. This means that $L_1$ is bright and $L_2$ and $L_3$ are dimmer than $L_1$ and equally bright.

(b) Because one branch of the parallel combination has been removed, the total resistance of the circuit increases. This means that the total current decreases. $L_1$ is therefore less bright. The current in $L_2$ is the same as that in $L_1$, so these two lamps now have the same brightness, which is greater than the original brightness of $L_2$.

**Commentary**:

(a) The explanation is given before the comparison but both parts of the answer are given.

(b) The basis of the answer is clearly stated – the total resistance increases – and its effect on the current and the brightness of the lamps.

## Data analysis questions

㉑ **Model answer**:

(a) Assuming that the uncertainty in the thickness measurement is ±0.05 mm, i.e. half the resolution, the percentage uncertainty in the thickness, $p_t$ is given by:

$p_t = \dfrac{0.05}{5.15} \times 100\% = 0.97\%$

The percentage uncertainties in the length and width, $p_l$ and $p_w$ are much smaller:

$p_l \approx \dfrac{0.01}{29.7} \times 100\% = 0.03\%$

and $p_w \approx \dfrac{0.01}{21.0} \times 100\% = 0.05\%$

So the total uncertainty in the volume will be totally dominated by the uncertainty in the thickness.

(b) Mean value of length = (29.72+29.71+29.71+ 29.70+29.71)/5 = 29.71 cm

Mean value of thickness = (21.03+21.05+21.05+ 21.04+21.04)/5 = 21.04 cm

∴ Volume of 500 sheets

$$= \text{length} \times \text{width} \times \text{thickness}$$
$$= 29.71 \times 21.04 \times 5.15 \text{ cm}^3$$
$$= 3219 \text{ cm}^3 \pm 0.97\%$$

Volume of 1 sheet $= 6.44 \text{ cm}^3 \pm 0.97\%$
$$= 6.44 \pm 0.06 \text{ cm}^3$$

**Commentary**:

(a) The resolution of the metre rule is stated (by implication) and the reasonable claim that the experimenter can read to half the resolution – this justifies the subsequent work. The percentage uncertainties of the thickness, length and width of the ream of paper are correctly calculated and compared.

The statement about the equality of the percentage uncertainties in the thickness of the block of paper and an individual sheet is not perhaps necessary – but is correct and doesn't hurt!

(b) The mean values of the length and thickness are correctly calculated and combined to give the volume of the block. The percentage uncertainty stated is that of the thickness alone as we have already established that this dominates.

The volume of a single sheet is correctly calculated by dividing by 500 and the percentage uncertainty is the same as that of the block.

The absolute uncertainty is correctly calculated and expressed to 1 s.f. The volume of one sheet is expressed to two decimal places as the uncertainty in volume is in the second decimal place.

㉒ **Model answer:**

(a) (i)

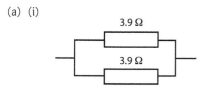

(ii)

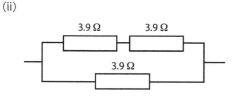

(b)

| $R/\Omega$ | $I/A$ | $\dfrac{1}{I/A}$ |
|---|---|---|
| 1.3 | 0.940 | 1.06 |
| 1.95 | 0.667 | 1.50 |
| 2.6 | 0.533 | 1.88 |
| 3.9 | 0.373 | 2.68 |
| 5.85 | 0.250 | 4.00 |

(c)

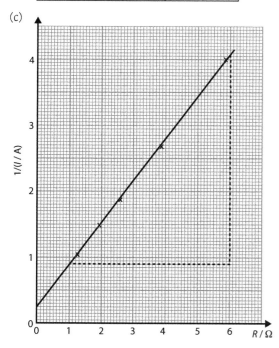

(d) The equation suggests that a graph of $1/I$ against $R$ should be a straight line with a positive gradient of $1/E$ and a positive intercept on the vertical axis of $r/E$. The graph is a straight line with a positive gradient., with the points showing very little scatter. The intercept on the $1/I$ axis is positive [+0.23 A$^{-1}$] so both aspects of the graph support the theoretical equation, with $E$ and $r$ being constant.

(e) The gradient of the graph $= \dfrac{(4.07 - 0.88)}{(6.00 - 1.00)}$

$= \dfrac{3.19}{5.00} = 0.638$

The gradient is $\dfrac{1}{E}$ $\therefore E = \dfrac{1}{0.640} = 1.57$ V

The intercept is 0.23 A$^{-1}$ and equal to $\dfrac{r}{E}$

so $r = 1.57 \times 0.23 = 0.36\ \Omega$

**Commentary:**

(a) The resistor arrangements give the required resistances. The question did not require the candidate to justify the answers by calculation. The 1.95 $\Omega$ parallel combination is easy to spot (the value is half 3.9 $\Omega$). The 2.6 $\Omega$ arrangement is more tricky but it is obviously the only combination of three 3.9 $\Omega$ resistors which would give a resistance between 1.95 $\Omega$ and 3.9 $\Omega$.

(b) The 1/current values are given to 3 figures, which agrees with the current data.

(c) The graph has labelled axes which exploit the full range of the graph paper. The scales are uniform. The units are given. The points are plotted within ½ a square. A reasonable line is drawn consistent with the data. The line is extended to meet the vertical axis – which is required later on.

(d) The various aspects of the line are commented on: straight, positive gradient and intercept. They are compared to the prediction. The degree of scatter is commented on, which is important as, if there had been a large degree of scatter, the strength of the confirmation of the relationship would have been lower. There is no requirement to state the value of the intercept here, but it is used in part (e).

(e) The gradient is determined using a large triangle on the graph – the triangle is drawn in, helping the examiner to see what the candidate has done. Note that it is not necessary to include units on the gradient.

The gradient and intercept are used correctly to determine $E$ and $r$.

㉓ **Model answer**

(a) (i)

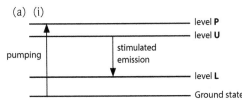

(ii) Photon energy,

$$E = \frac{hc}{\lambda} = \frac{6.63 \times 10^{-34}\ \text{J s} \times 3.00 \times 10^{8}\ \text{m s}^{-1}}{1.05 \times 10^{-6}\ \text{m}}$$

$$= 1.89 \times 10^{-19}\ \text{J}$$

Level U $= 1.89 \times 10^{-19} + 0.3 \times 10^{-19}$

$$= 2.2 \times 10^{-19}\ \text{J} \qquad \text{(2 s.f.)}$$

(b) A population inversion is achieved between levels U and L as follows: Electrons are raised from the ground state to the very short-lived level P by a pumping mechanism (e.g. optical pumping). They drop down very quickly to the metastable level U. Level L is a very short lived state, so any electrons which drop down to level L almost immediately drop to the ground state. Hence the population of level U is greater than level L.

When a photon from the U → L transition, e.g. from a spontaneous emission, is incident upon an atom in level U it stimulates the transition to level U by the emission of a second photon of the same energy travelling in the same direction. Thus there are now two photons instead of one, which can stimulate further emission. This is a cumulative process in which the number of photons increases exponentially.

**Commentary:**

(a) (i) The pumping and stimulated emission transitions are clearly labelled.

(ii) The photon energy is correctly calculated. The answer clearly identifies that the level U energy is the photon energy above L and therefore level U is the photon energy plus $0.3 \times 10^{-19}$ J above the ground state. The answer is given to 1 d.p. because the energy of level L is only given to 1 d.p.

(b) There are two parts to this answer. One part deals with how a population inversion is obtained and the other explains how this enables light amplification through stimulated emission.

In the population inversion it is important to deal with the pumping and the lifetime of the three states P, U and L. All these are highlighted.

The light amplification section of the answer indicates that the stimulating photon needs to have the same energy as that of the population inversion energy states and that both resulting photons can go on to stimulate more emission, identifying this with light amplification. The question does not require mention of the phase of the photons and the coherence of the light beam.

# Index